LOGICAL DESIGN
OF AUTOMATION SYSTEMS

SANDER B. FRIEDMAN
Professor Emeritus, Manufacturing Engineering
Miami University, Oxford OH

PRENTICE HALL, Englewood Cliffs, New Jersey 07632

Library of Congress Cataloging-in-Publication Data

Friedman, Sander B.,
  Logical design of automation systems / Sander B. Friedman.
    p.   cm.
  Includes bibliographical references.
  ISBN 0-13-540097-X
  1. Automatic control.   2. Logic design.   I. Title.
TJ211.F75 1990
670.42'7--dc20
                                                          89-37658
                                                              CIP

Editorial/production supervision
  and interior design: Mary Kathryn Leclercq
Cover design: 20/20 Services, Inc.
Manufacturing buyer: Denise Duggan

The author and publisher of this book have used their best efforts in preparing this book. These efforts include the development, research, and testing of the theories and programs to determine their effectiveness. The author and publisher make no warranty of any kind, expressed or implied, with regard to these programs or the documentation contained in this book. The author and publisher shall not be liable in any event for incidental or consequential damages in connection with, or arising out of, the furnishing, performance, or use of these programs.

© 1990 by Prentice-Hall, Inc.
A Division of Simon & Schuster
Englewood Cliffs, New Jersey 07632

All rights reserved. No part of this book may be
reproduced, in any form or by any means,
without permission in writing from the publisher.

Printed in the United States of America
10 9 8 7 6 5 4 3 2 1

ISBN   0-13-540097-X

Prentice-Hall International (UK) Limited, *London*
Prentice-Hall of Australia Pty. Limited, *Sydney*
Prentice-Hall Canada Inc., *Toronto*
Prentice-Hall Hispanoamericana, S.A., *Mexico*
Prentice-Hall of India Private Limited, *New Delhi*
Prentice-Hall of Japan, Inc., *Tokyo*
Simon & Schuster Asia Pte. Ltd., *Singapore*
Editora Prentice-Hall do Brasil, Ltda., *Rio de Janeiro*

Contents

PREFACE   vii

   Acknowledgments   ix

Chapter **1**
INTRODUCTION TO AUTOMATION   1

    1.1   Automation Economics: General   3

    1.2   Cost Components   3
        1.2.1   Design Considerations
        1.2.2   Fabrication Considerations
        1.2.3   Operating Considerations

    1.3   Costing Procedures   5

Chapter **2**
INTRODUCTION TO AUTOMATION DESIGN   9

    2.1   Historical Background   10

    2.2   Concept of a Binary Variable   11

    2.3   Concept of Logical Synthesis   12

    2.4   Industrial Automation Schemes   14

    2.5   Problems   15

Chapter 3
BOOLEAN ALGEBRA 17

    3.1 The Elements 18

    3.2 The Relationships 18

    3.3 Basic Operators 18
        3.3.1 NOT
        3.3.2 AND/NAND
        3.3.3 OR/NOR, XOR/XNOR
        3.3.4 MEMORY Function
        3.3.5 Hierarchy of Operations

    3.4 Postulates and Theorems 24
        3.4.1 Postulates of Boolean Algebra
        3.4.2 Theorems of Boolean Algebra

    3.5 Theorem Proofs 25
        3.5.1 Proof Using Truth Tables
        3.5.2 Proof Using Theorems

    3.6 Definitions 27

    3.7 Equation Manipulations and Reductions 28
        3.7.1 The Complement of an Expression
        3.7.2 The Dual of an Expression
        3.7.3 The Universal Operators

    3.8 Problems 31

Chapter 4
EQUATION OPTIMIZATION METHODS 33

    4.1 Algebraic Method 34

    4.2 Mapping Method 36

    4.3 Simplification Using Prime Implicants 43
        4.3.1 Definitions and Proofs
        4.3.2 Coding
        4.3.3 Derivation of the Prime Implicants
        4.3.4 Forming the Optimal Expression
        4.3.5 Functions Expressed Non-Canonically
        4.3.6 Multiple Output Systems

    4.4 Summary 51

    4.5 Problems 52

Chapter 5
IMPLEMENTATION OF LOGIC FUNCTIONS 55

    5.1 Logic Implementation 56

    5.2 Mechanical Input, Mechanical Output Systems 57

    5.3 Mechanical Input, Electrical Output Systems 58

    5.4 Electrical Input, Electrical Output Systems 59

Contents v

    5.5 Fluid Power Systems 62

    5.6 Fluidic and Electronic Systems 68

    5.7 Other Devices 70

    5.8 Programmable Logic Controllers 71
          5.8.1 Background
          5.8.2 Comparisons to Other Hardware Systems
          5.8.3 Programming PLCs

    5.9 Selection of Implementing Hardware: Other Criteria 77

    5.10 Problems 79

Chapter 6
COMBINATIONAL LOGIC SYSTEMS 81

    6.1 Logic Design Symbology 82

    6.2 Typical Systems 83
          6.2.1 Combinational System Hazards
          6.2.2 Hazard-Free Circuits

    6.3 Problems 99

Chapter 7
PROBLEM STATEMENTS 103

    7.1 Statement of Combinational Problems 104

    7.2 Statement of Sequential Problems: Extended Mapping Technique 105

    7.3 Statement of Sequential Problems: Primitive Flow Tables 109

    7.4 Statement of Sequential Problems: Logic Specification Charts 111

    7.5 Simplification of Logic Specification Charts 115

    7.6 Problems 121

Chapter 8
SEQUENTIAL LOGIC SYSTEMS: COUNTING AND STEPPING METHODS 125

    8.1 Theoretical Basis 126

    8.2 Counting Systems 128
          8.2.1 Binary Counting Systems

    8.3 Rotary Steppers 131
          8.3.1 Rotary Stepping Switches
          8.3.2 Rotary Solenoids
          8.3.3 Drum Steppers

    8.4 Examples 137

    8.5 Problems 141

Chapter 9
SEQUENTIAL LOGIC SYSTEMS: PASSIVE MEMORY TECHNIQUES 143

    9.1 Passive Memory Hardware   144

    9.2 Stochastic Systems   148

    9.3 Deterministic Systems   151

    9.4 Problems   154

Chapter 10
SEQUENTIAL LOGIC SYSTEMS: ACTIVE MEMORY TECHNIQUE   157

    10.1 Active Memory Hardware   158

    10.2 Active Memory Design Techniques   161
        10.2.1 First Steps
        10.2.2 Operational Flow Charts
        10.2.3 Excitation Charts

    10.3 Output Statements   179

    10.4 Sequential Hazards   181

    10.5 Summary   182

    10.6 Problems   183

Chapter 11
AUTOMATION CASE STUDY   185

    11.1 Introduction   186

    11.2 Initial Considerations   186

    11.3 Initial Analysis   186

    11.4 Problem Statement   188

    11.5 Design of the Control System   190

    11.6 Results   192

Chapter 12
COMPUTER ASSISTED DESIGN   196

    12.1 PRIMP   196

    12.2 TRUTAB   199

SELECTED BIBLIOGRAPHY   201

LISTING OF COMPUTER PROGRAMS   204

INDEX   211

Preface

Before dealing with the design of industrial automation systems, it is appropriate to consider what automation is and how it has developed over the years.

Automation is defined in the *Oxford Dictionary of the English Language* as coming from the Greek root meaning "that which happens of itself." *Mechanize* means "to make or render mechanical," that is, to substitute machine effort for human effort. *Logic* is defined in the same volume as ". . . the science or art of reasoning as applied to some particular department of knowledge or investigation."

The design of automation systems applies logical principles, implemented through various devices such as valves, relays, switches, solid state devices, and the like, to control the sequencing of industrial work performing mechanized devices.

Automation systems should not be confused with the automatons of the sixteenth and seventeenth centuries, those remarkably complex, lifelike toys that performed incredibly ingenious feats. Unfortunately, they did no useful work and therefore were only curiosities to amuse their viewers.

It is reasonable to think of the Industrial Revolution in England as being the real beginning of industrial mechanization, although isolated examples of mechanical devices existed for many years prior to the onset of that epoch. Full mechanization of the manufactory (factory) was not truly realized until the early nineteenth century in the Connecticut valley of the United States. This led to the "American Method of Manufacturing" which used interchangeable assembly of parts manufactured on relatively inexpensive machines.

The first practical automation scheme was implemented in the late eighteenth century in the weaving industry in England. This was done on the Jacquard loom which used a wooden board into which a series of holes had been fabricated. These holes enabled rods, or fingers, to pass through the board if a hole was directly below the finger. If there was no hole, the finger remained in its undisplaced position. The position of the fingers in turn controlled the pattern that was woven on the loom.

This development truly automated a mechanized process by using logical methods in which an act either was or was not performed; that is, a two-state logic system was applied.

There is an inherent problem with automation, however: these systems are expensive to design, build, and maintain. Applications of automation were therefore limited to the repetitive production of large quantities of identical parts, where the large investment in controls and machinery could be distributed over many items. Even with this additional cost, the economies of fabrication so outweighed the additional cost of tools and equipment that one could not afford to proceed in any other manner.

Because of the above, industrial manufacturing was effectively divided into two categories:

- Large quantities of product where the process was capital intensive, and
- Smaller quantities that were produced on unautomated machines operated by skilled or semi-skilled personnel resulting in a labor intensive process.

The rule became and still is: *If the quantity and regularity of production will justify it, automate the process*!

Modern logical techniques applied to the manufacturing processes permit the *selective* automation of these processes with much lower capital investment than that required by *total* automation. This in turn means that significantly smaller quantities can be used to justify automation, thereby reducing overall cost and enabling industry to remain and/or become competitive.

It is the purpose of this work to describe the tools and techniques used in the design of automation systems in a logical and orderly fashion, so that the reader will be able to apply these techniques in an economically viable and responsible way.

ACKNOWLEDGMENTS

It is only proper that prior work in this field be acknowledged. All logical analysis is based on the mid-nineteenth century works of George Boole. His books written in 1847 and 1854 have been republished by Oxford University Press, Basil Blackwell of Oxford, and Dover of New York. Other references are given in the bibliography, and are indicated by numbers in parentheses after the reference's name, such as (93). Particular attention should be given to the works of Shannon, Fitch, and Martin. Special thanks are given to Professors Hugh R. Martin and R. L. Woods for their contributions in fluid logic and its applications, as well as their support and suggestions.

S.B.F.

Chapter 1

INTRODUCTION TO AUTOMATION

There is never a question about whether a manufacturing process should be automated: It should be. Rather, the question is: To what extent should the process be automated?

There are always many questions inherent in this decision process. Is it desirable to mechanize the process, that is to substitute machine power for human effort? Should the process be automated by completely eliminating human intelligence and control from the system, that is mechanize as well as substitute machine *intelligence* and control? Should there be no change to the existing process (the null hypothesis)? Or is the best answer contained somewhere within the extremes?

The constraints that effect the automation decision process are, in order of importance:

- Technological constraints,
- Economic constraints, and
- Fiscal constraints

The technological constraints affect both the power or work producing system and the control system. It could be technologically ludicrous to attempt to automate some systems with "Rube Goldberg mechanisms." However, short of this reduction to the absurd, the vast majority of industrial processes could be profitably automated.

There are many alternate automation hardware systems, all of which have their place under specific application conditions. Systems such as fluid power, fluidic, electronic, electromechanical, microprocessor control, and the like, all have specific applications where each could be considered *best*. The choice of the automation system hardware is discussed at much greater length later.

The technological constraints that should be considered when selecting a logic hardware system are:

- Function and Compatibility,
- Reliability and maintainability, and
- Cost

These, in conjunction with the economic and fiscal constraints, should result in the optimal control system for a given application. In the final analysis, the system selected should reliably do exactly what is required of it, with no unnecessary features, for as long as is required, at the lowest total cost.

Automation has no preference concerning how it is implemented. Sequential control techniques that have been well developed for electronic and telephonic systems can be applied to other hardware systems. One should consider the fact that fluid and electromechanical logic hardware perform some very basic, and often unique, logic functions compared to electronic and fluidic systems. Power levels of fluid and electromechanical systems are higher, both for sensing and computational functions. This can result in systems that are essentially impervious to spurious signals (noise) resulting in significantly higher levels of reliability. The price that is paid is in size, component cost, and speed of operation. For these reasons, it is essential that the control function be accomplished with an absolute minimum number of components.

Considering the special usage requirements and logic functions of some hardware systems, it becomes clear that unique component capabilities must be considered throughout the design process, including synthesis and implementation. Each of the synthesis procedures discussed herein is better suited to the most flexible solution; therefore the hardware considerations are not delayed until the implementation step.

The fiscal constraint may be simply stated as the requirement that adequate capital must be available, either as cash on hand or as a line of credit, to make the required investment in the desired automation system.

An extended discussion of the economic constraints follows.

1.1 AUTOMATION ECONOMICS: GENERAL

In any design, the first activity is to define the problem to be solved both its nature and extent, and to assure that there is in fact a need for the solution; then the various parameters and constraints must be established. All too often in the design of an automation system, the economic constraints are considered to be of secondary importance. They are not even recognized until the last steps of the project since designers intuitively consider their activity to be the most costly component of the project. In addition, the nature of the physical system is essentially unique resulting in a single system being required.

The purpose of this discussion is to realistically define the factors affecting an automation system design project in addition to the technological considerations and to suggest a rational design procedure to follow.

The driving consideration in any engineering design is the concept that the system being considered must meet its design objective. Within that constraint, the optimal system should do exactly as much as is required for a specified service life, at the lowest overall cost.

1.2 COST COMPONENTS

The components of overall cost for an automation system are:

- Design cost including the feasibility study, the determination of the appropriate economic level of automation/control, and the actual engineering and design of the physical system,

- Component, assembly, and testing costs, and
- Operating, reliability, and maintenance costs for the service life of the system.

1.2.1 Design Considerations

Assuming that the need has been established for an automation system, the first consideration should be the extent of the automation. It is not appropriate to choose hardware at this time, but rather only to consider the logic of control. Once the control logic is specified, then the most appropriate implementation can be selected.

The extent of the automation is a purely economic decision. It can be safely generalized that three-quarters of full automation benefits can be accomplished for one-quarter of full system cost. Using this concept as a guide, one can establish full automation system parameters; then design that system using the techniques of combinational and sequential logic. Some of these design techniques have been computerized, but whether or not they are available, it is still presupposed that, at the least, the individuals using these tools are competent to accurately specify appropriate systems.

Upon completion of the design, the system can be analyzed for cost effectiveness, and reasonable cutoff points established, based on such considerations as overall quantities, production quantities, delivery schedules, material availability, available personnel, and so forth.

At this point, the nature of the input-output interface should be investigated in order to determine the optimal hardware system. Again, the parameters should be function, cost, reliability, and maintainability.

1.2.2 Fabrication Considerations

It is not reasonable to assume a clear-cut advantage in total cost considerations for any one hardware system for all applications. It is reasonable, however, that in any given industry or situation, total fabrication and testing costs for one particular hardware system will be more attractive than its alternatives. This is due to a combination of interface costs, acquisition costs, fabrication and assembly costs, and testing costs.

Any given industry will be competent in specific but limited areas due to both the nature of the industry and the products with which it is involved. Fabricators and assemblers will be more familiar with their everyday processes rather than new, different, or exotic techniques. Selection of the appropriate hardware is essentially a function of the above considerations coupled with reliability and maintainability factors. The relative effective cost of any system can best be expressed as the sum of the conventional cost factors and the lost-opportunity costs resulting from either faulty operation of the system, or in the pathological case, actual failure and breakdown.

1.2.3 Operating Considerations

Two tautologies can best express this category of cost:

- One pays for design and fabrication once; one pays operating costs for the service life of the system.
- A machine is potentially profitable only while it is producing a product.

Historically, the costs of energy were so insignificant that they were normally ignored in most design considerations. There was no clear cut advantage to the use of electric, pneumatic, hydraulic, or any other source of power. This is still generally true, but while the cost for energy is no longer insignificant and must be considered in any total cost evaluation, there is probably still little cost advantage of one source over another.

Of much greater significance are the maintenance and lost-opportunity costs. With the exception of some of the largest and most exotic high-tech industries, the expertise of most maintenance organizations is limited to mechanical rebuilding; and electrical, hydraulic, and pneumatic repairs. For example, it is unreasonable to expect that a fully competent mechanical millwright would also be expert in servicing electronic systems.

When one considers the combined costs for all categories coupled with the lost-opportunity costs incurred when product is not being produced, it becomes evident that the most important considerations in the design and construction of an automated system are the degree of automation that can be justified, and the personnel skills available.

1.3 COSTING PROCEDURES: JUSTIFICATION AND OPERATING COSTS

There are two different techniques required for economic evaluation of an automation project. One is used when the acquisition of new capital equipment is to be justified. The second is appropriate when considering the operating cost of the automatic machine or, more accurately, the overall cost of the product. These techniques are fully discussed in many references and will not be considered here, other than to say that all evaluations must be made on a total economic life basis for each option under consideration.

What is of concern is the conventional distribution of costs on a direct and indirect basis.

The essence of mechanization and automation is the removal of human effort and intelligence, respectively, from the repetitive phases of the manufacturing process. It therefore seems counterproductive to base the cost for an automated process on the labor content of the process. This is especially true when one considers that even in the case of conventional labor intensive manufacturing, the actual direct cost of labor is only a small portion of the total costs that are incurred.

For the purposes of this discussion, all manufacturing processes will be considered as ranging from labor-intensive to capital-intensive. In the former, the dominant cost component is the direct-plus-fringes labor cost. In the latter, the distributed hourly cost of the equipment, make-ready costs, setup and tooling, and manufacturing engineering are the dominant costs. The problem is that it is necessary to compare labor-intensive processes with capital-intensive processes. Therefore, a rational and equitable cost base must be established.

In conventional labor-based cost analysis, the direct cost consists of labor-plus-fringes costs for the manufacture of an item. All other costs, both fixed and variable, are then added onto the direct costs in the form of burden or overhead.

In order to fairly compare labor and capital intensive processes, it is essential that the distribution of burden be reevaluated. Those items that reflect the capital costs that are directly attributed to the cost center must be excluded from burden and considered as additional direct costs. This technique is essential for a capital-intensive process, and is not only fair but is actually sensible for labor intensive processes. It is more complicated, but with the advent of computer based accounting systems, the differences are insignificant.

The process of adjusting overhead charges is known as budgeting. The relevant budget items that are normally contained in overhead for distribution over a plant wide base must be deleted and their costs directly charged to the appropriate automated work or cost center. The charges referred to consist of, but are not limited to, space charges, power charges, support and engineering charges, and equipment charges.

The space charges of concern include rent, maintenance, repair, and any other occupancy charges encumbered essentially due to the presence of the equipment. Similarly, the cost of power not only for the operation of the machine, but also for all the support equipment, must be considered.

In any automated process, using either a dedicated controller or computer assisted manufacture, a considerable engineering effort is required both in the initial make-ready period when the machine is first coming on line, as well as during the operating life of the machine. The differences in support levels for automated processes are significantly greater than for conventional processes. By not separating these charges from burden, an essentially unfair load is placed on the conventional work center resulting in the automated work center appearing more desirable that it really is, while the unautomated station appears more costly.

The final major category to be considered is equipment cost. This includes depreciation, installation, dedicated general-purpose tooling, repair, maintenance, outside services, and any other cost that may be appropriate. With the costs of automated equipment what they are, this category could be many times greater than the conventional direct costs. As with engineering and support costs, the result of not transferring these items from indirect to direct charges is to make the automated system appear more desirable.

This method of distributing costs from overhead to the work center can be considered as an augmented or extended man-machine costing system. Specific techniques of budgeting are discussed in many cost accounting and cost estimating references such as Ostwald.[28]

Chapter 2

INTRODUCTION TO AUTOMATION DESIGN

2.1 HISTORICAL BACKGROUND

The foundations of formal logic were developed during the fourth century B.C. by the Greek philosopher Aristotle. His work on the laws of logical reasoning was so complete that only minor adornments were added through the thirteenth century A.D. by such thinkers as Abelard, Occam, John, and Sherwood. Since that time, formal or Aristotelian logic has remained essentially unchanged.

The basic statement of Aristotelian logic is:

> A statement is either true or it is false;
> it cannot be both, it cannot be neither.

Many philosophers tried without success to devise a suitable mathematical representation of the sentence based logical reasoning process of Aristotle. This was successfully done in 1854 when an English mathematician, George Boole,[3] devised a two-valued algebra that could be used in the modelling of true-false propositions. Since that time, Boolean algebra has been a favorite topic of mathematicians.

In developing his algebra, Boole realized that mathematical variables could be used to represent the truth value of verbal statements. For example, let the variable A represent the truth value of the statement "The contacts of the electrical switch are touching each other." According to the precepts of logic, this statement must be either true or false. If the statement is true, then the truth value of A is 1, or $A = 1$. If, however, the statement is false, then the truth value of A is 0, or $A = 0$.

Although it may appear obvious that switch contacts are either touching or not, it can reasonably be asked "How can one be sure that this statement is an accurate representation of fact?"

For any physical event or series of events, an experiment can be designed to determine the accuracy of the assertion of falsity; in order for a statement to be true, it must always be true. In other words, upon analysis of experimental data, if the results are ever false, the statement is false. In order to prove truth it is necessary to know that all outcomes, present, past, and future, are true. This is physically unrealizable, since complete and perfect knowledge of all outcomes is an unattainable goal. Therefore, where falsity can be proven, truth can only be implied.

In the case of logical automation systems, even though the truth of a statement can only be implied, it can be accepted as being extremely probable. If falsity develops in a relationship that has been assumed to be true, then it is said that a "malfunction" has occurred and appropriate remedial action is required, which may consist of doing nothing until the definition of truth can be reasonably changed.

It should also be noted that in cases of logical reasoning, equalities can be established through the use of the conditional statement: "if . . . , then

. . . ." This concept of mathematical representation of logical statements has led to the vast knowledge known as philosophical and mathematical logic and is of proven use in the process of logical analysis and control.

The next step in logical development is to let the variable A represent the state of the switch itself rather than a statement about the switch. Thus $A = 1$ if it is true that the switch contacts are closed, and $A = 0$ otherwise. In 1938, C. E. Shannon[34] used this reasoning and developed a technique whereby Boolean algebra could be used to represent electrical telephonic circuits using switches and relays. This was a critical development in logical design and gained immediate recognition and success.

The concept of mathematically representing logical statements was extended from combinational (simultaneous) logic systems to sequential circuit design requiring memory, by D. A. Huffman[19] and E. F. Moore[27] in 1954. Many other investigators became involved in the application of mathematical analysis to the design of complex switching and automation systems, leading to a body of knowledge known as *switching circuit theory*. E. C. Fitch[12-14] was a pioneer in applying the formal techniques of switching circuit theory to fluid power control systems, while others such as H. R. Martin and S. B. Friedman[15,16] were applying the techniques to industrial automation, and R. L. Woods,[36,37] G. E. Maroney,[22,23] and others were working in the area of fluid logic systems.

The increasing capability of electronic logic and coding, and the advent of low-cost programmable microprocessors and minicomputers, has led to present day industrial automation and computer assisted manufacturing systems.

2.2 CONCEPT OF A BINARY VARIABLE

The concept of letting mathematical variables represent the state of physical events is not limited to electrical switches and relays. It is also applicable to any physical event in which the main characteristic of interest is exhibited in either one or the other of two *exclusive* states or conditions. Examples include fluid power spool valves, semiconductor switching devices, coanda effect fluidic amplifiers, electrical relays and switches, toggle mechanisms, and spring loaded mechanisms. Thus, any device that exhibits this two-valued behavior is called a *binary device*. The state of the device can be represented mathematically by a binary signal or binary variable. Examples of binary variables include voltage present or absent, pneumatic line pressurized or vented, and mechanical linkage displaced or at rest.

Binary variables can be used to represent uncontrolled or independent outputs such as those coming from sensors, transducers, or other primary sources. These outputs can be considered as the *input variables* that in turn determine the system action. The response variables are termed the *output variables* since they are determined from the inputs and perform the system responses and/or actuations. Thus, binary variables can be used for both inputs and outputs.

2.3 CONCEPTS OF LOGICAL SYNTHESIS

There are four identifiable steps in the process of building a control system:

- System definition or a word description of the possible input combinations and the desired outputs resulting from the appropriate inputs,
- Synthesis of a mathematical model of the controller that will accomplish the task defined in the first step,
- Implementation of the above model using the appropriate physical hardware, and
- Analysis of the operation of the hardware system to insure that the resulting system will perform its design mission in an optimal manner.

The system definition step is more or less specified by the task at hand. It does require the engineer's judgment in some cases to fully describe what system response (output) is to be taken for the various input conditions.

Figure 2.1 illustrates the synthesis problem. Once the system definition or word description step is complete, the designer knows what inputs to expect and what output responses are desired. The task is to synthesize a logic control or automation system that has these required input-output relationships.

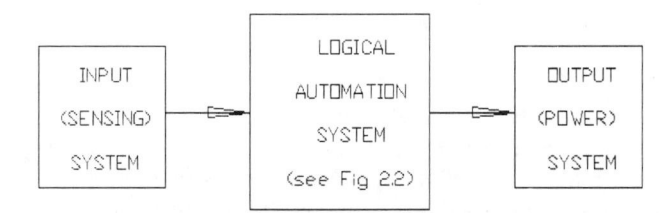

Figure 2.1 A Logically Controlled Automation System

The synthesis step is the significant contribution made by logical techniques. Before the development of logical synthesis, system designers had to more-or-less start with valves, relays, switches, and the like, and try to devise a way to interconnect them to obtain a working system. This was an intuitive process at best, and was more of an art than an exact science. Note that the intuitive designer would have to think in terms of interconnections of hardware devices and of the effect that one device had on another, that is "If I use this valve as a sensor and connect this port to this other valve,

Sec. 2.3 Concepts of Logical Synthesis

then fluid pressure will go from this line to move this other valve which in turn will move this cylinder. . . . This is one step removed from the word statement which is really the problem that one is trying to solve.

With logical synthesis procedures, the designer is able to convert the word statements into a mathematical or structural format and, by logical techniques, convert this format into binary equations representing the desired system response. Then the binary equations are implemented with hardware selected from an inventory of devices; that is, a table that indicates which device will perform the various desired logic functions. With this synthesis procedure, it is possible to obtain a system that will perform the desired function without ever having to consider the effects of hardware interconnections as is required when doing intuitive design.

Using this logical synthesis approach, it is possible to input the word description of the desired operation, in coded form, to a computer. Using the appropriate software and the problem statement, the design of the required control system can be computer generated.

Figure 2.2 shows the relationships of the various branches of logical design. Logical synthesis is defined as the process of obtaining a mathematical representation (using Boolean algebra) of a binary control system that will fulfill the requirements of the word statement.

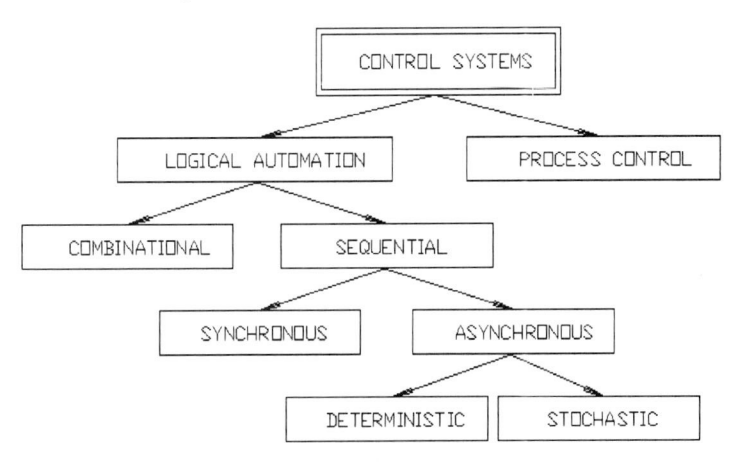

Figure 2.2 Logical System Structure

There are two basic branches of logical synthesis: combinational and sequential. Combinational systems are characterized by the fact that the output is dependent upon the present inputs *only* and does not depend on the sequence of the input excitations. In sequential systems, the output is a function not only of the present value of the inputs, but also of the previous values or sequence of the input excitations. This obviously requires that the logical solution of a sequential problem have a memory capability, whereas a combinational system, being time independent, requires no memory.

With the increasingly complicated operations that could be occurring in a sequential system, time synchronization could become necessary, resulting in a further breakdown of sequential systems into asynchronous and synchronous operation. Asynchronous systems are characterized by the fact that system outputs are generated as soon as the appropriate input combinations appear; in other words, they are *event* timed. In synchronous systems, all logic outputs and transfers occur at specified intervals of time. The intervals of time are determined by a clock pulse. Typically, all the inputs are formed and are awaiting the clock pulse before the output or transfer can occur. Synchronous systems are therefore said to be *clock* timed.

Since synchronous systems require a clocked timing signal, they require an additional input. Their timing must be based on the longest possible duration of all steps in the process. Since they are function insensitive, that is clock timed as described above, they can result in operationally inoptimal system automation and should be used only in the case where they can be economically justified. For these reasons, the primary concern of this work is with combinational and asynchronous sequential automation and control systems.

Finally, sequential logic systems can be either deterministic, where the system is to always operate with the same sequence of input and output steps; or stochastic, where the inputs operate in a random pattern such as in a sorting or selection system.

2.4 INDUSTRIAL AUTOMATION SCHEMES

Further generalizations of logic have eventually led to the general purpose digital computer as we now know it. Recent advancements have resulted in the microprocessor, the microcomputer, and the programmable controller; all of which are available at very modest cost so that direct digital control with a general purpose computer is now as feasible as special purpose logic control systems. As a consequence, the basic synthesis procedure for sequential automation might appear to have outlived its usefulness due to the development of more advanced and sophisticated hardware systems. Furthermore, the concern for minimum logic hardware for implementation may not be as critical since electronic logic elements are so inexpensive.

Since some of the basic synthesis procedures, as well as the concern for minimum hardware, are becoming less important in the electronic domain, it would be reasonable to inquire as to the appropriateness of logical synthesis.

The answer to the basic question "Are alternative automation hardware systems viable?" is "Yes, maybe!" depending on many technical considerations.

For example: In some cases, electrical power is highly undesirable due

Sec. 2.5   Problems 15

to the possibility of shock, explosion, or other environmental hazards. This is especially true in any industry where the product, the materials used in the process, or the process itself present a *clear and immediate* hazard to the personnel concerned with production.

In other cases, the presence of alternating current as an electrical power source may generate significant levels of electrical noise. The use of low-level electronic microprocessors and controllers may be impractical due to the system's inability to distinguish between a viable signal and a spurious one.

2.5 PROBLEMS

The following problems are intended to introduce the reader to the intuitive design process. Each problem should be analyzed for the appropriateness of the inputs, the required input combinations, the necessary control outputs, and the system outputs, as required. Do *not* attempt to use the synthesis techniques developed in subsequent chapters. Rather, rely on your own logical thinking process and good judgment. *Note that these problems will be reconsidered after the appropriate synthesis methods have been discussed.*

1. Design and draw an electrical switch system that will perform the following functions.

    When switch A and switch B are ON, device Z_1 is to be activated. If switch A is ON and switch C is OFF then device Z_2 is to be energized. Z_2 should be activated in the event that switch A is held ON for a significant period of time; however, it should never come ON if switch B is ON. Provide for the possibility of operating Z_1 if switch C is ON.

2. Design and draw an electrical switch and relay system that will perform the following decisional control functions.

    Three different sized boxes are to be sorted. Their dimensions vary significantly, so minor variations can be ignored. The sorting is to separate small, medium, and large boxes. The boxes are transported on a conveyor. At the end of the conveyor are three bins, one for each sized box. The boxes are not to be dropped or otherwise mishandled.

3. Design and draw an air valve system consisting of ON/OFF valves, and a device which shall be called a *MEMORY* which recalls which of its two inputs was ON last. This system is to provide control to automate the following. Use as few components as possible.

    A pneumatic press is to be modified such that operators are prevented from endangering themselves by placing their hands in the work area while the machine is functioning. The system requirements are as follows:

    - Two operating valves are to be used, one each for the right and left hands.
    - The press ram can move downward if-and-only-if both operating valves are operated.
    - The press cannot be recycled unless both operating valves are OFF prior to the next stroke.

Chapter 3

BOOLEAN ALGEBRA

An abstract algebra has a set of elements, a set of operators, and a set of relationships that make up a system. Boolean algebra is a classic example of an abstract algebra which can be used to symbolically represent and manipulate concepts. Automation system design is dependent on a limited subset of the complete Boolean algebra; therefore, only those aspects that are useful in the design of automation systems will be discussed here. For more information in depth, the reader is directed to any one of the many fine sources in the reference list, such as (17), (20), (25), (35).

3.1 THE ELEMENTS

In Boolean algebra, one is concerned with interrelationships among variables. These variables are binary since they can have one and only one of two states. The values of these states will be designated as 0 or 1, which will be used to represent false or true, open or closed, unpressurized or pressurized, voltage absent or voltage present, and so forth. The variables themselves will be named or noted by letters such as $A, B, C, \ldots, X, Y,$ and Z.

In a scalar (or ordinary) algebra, a variable can be in any one of an infinite number of states, ranging from minus infinity to plus infinity. However, since Boolean algebra is binary, each variable is permitted to have one and only one of the two states described above.

3.2 THE RELATIONSHIPS

In a scalar algebra, there are many relationships among expressions that are commonly used, including equality, less than, greater than, and so on. In automation, the Boolean relationship that is of primary importance is equality. Two Boolean expressions are said to be equal if they have the same output values for *all* combinations of the input variables. Equality of two expressions will be designated by the customary equal sign " = ".

3.3 BASIC OPERATORS

In a scalar algebra, there are many operators such as inversion, addition, subtraction, multiplication, and division. Some operators involve only one variable, while others may involve two or more at one time. In scalar algebra, the operations are defined or specified either by a word statement or by a table illustrating the operation process.

Sec. 3.3   Basic Operators

An example of an operation on only one variable is *inversion*. The operation *multiplicative inversion* of a number is called *division*, or 1 divided by the number. Similarly, *additive inversion* is called *subtraction*, or the change of sign or sense of a number.

In a scalar algebra, there are two operators that involve combinations of two or more variables: *addition* and *multiplication*. These operators are defined by addition and multiplication tables. Addition and multiplication tables list combinations of the independent variables and specify the value of the dependent variable for the operation. In effect, addition and multiplication tables are an attempt at an exhaustive listing of all combinations of the values of the independent variables. Of course, an exhaustive list of different combinations of numbers between plus and minus infinity is in itself infinitely large and therefore unmanageable. For this reason, values in the tables are proven by implication, i.e. if a statement is true for the nth occurrence of a combination of variables, and it is also true for the $(n + 1)^{st}$ occurrence, then it is considered that truth has been *implied* for all occurrences. This eliminates the need for infinitely large tables.

In Boolean algebra, operations involving one or two variables are defined by word descriptions and tables; these tables are called *Truth Tables*. A truth table is, by definition, a perfect and exhaustive listing of all possible combinations of the independent variables (input), and the corresponding truth values of the dependent variables (outputs).

3.3.1 NOT

An operation on a single variable in Boolean algebra is the operation NOT or COMPLEMENT. This operator is defined such that the value of the COMPLEMENT of a variable is the other value of the variable. For example; if $A = 1$, then NOT $A = 0$; and if $A = 0$, then NOT $A = 1$. The operator NOT is designated by an apostrophe " ' " immediately after the variable or negated term, or a bar over the variable or function. While both techniques are in use, the prime or apostrophe will generally be used throughout this work. In specific cases, the overbar will be used just to illustrate its application. The NOT operator is shown in the following truth table.[1]

A	Z = A'
0	1
1	0

Figure 3.1   The Operator NOT

---

[1] In the truth tables in this section, A and B are used as the inputs (independent variables) and Z as the output (dependent variable).

3.3.2 AND/NAND

The operators that involve at least two variables[2] in Boolean algebra are the operators AND and OR. The operator AND is defined as being true if and only if all the independent input variables are true, and false for all other input conditions. This operator is shown in the next truth table. The operator AND is normally indicated by a dot between variables, or juxtaposition of the variables as multiplication is in scalar algebra[3].

A	B	Z = AB
0	0	0
1	0	0
1	1	1
0	1	0

Figure 3.2 The Operator AND

A further definition of an operator that can be defined as a function of at least two variables is the operator NAND, which is a combination of NOT and AND, or the complement of AND; thus the contraction of the two words. In performing the NAND function, the AND operation is carried out first; then the result is complemented. The contraction AND-NOT would be a much better description of the operator but this acronym is not as satisfactory. The NAND function is illustrated in Figure 3.3. When comparing the truth tables for AND and NAND, it can easily be seen that they are the complements of each other.

A	B	Z = (AB)'
0	0	1
1	0	1
1	1	0
0	1	1

Figure 3.3 The Operator NAND

3.3.3 OR/NOR, XOR/XNOR

The other operator that involves combinations of two or more variables is OR. OR is said to be true if any input variable, or combination of variables, is true. This is called the *inclusive* OR, to distinguish it from XOR or the

---

[2] A similar truth table could be generated for three independent variables. Since this is the case, and using the law of implications, if truth is shown for two variables, it is also shown for any number of variables greater than two.

[3] The choice of logic symbols used to indicate the operators is somewhat unfortunate since they are identical to those used in scalar operations. However, since this is the *standard* symbology, it will be followed. The reader is cautioned against confusing scalar and logical operators.

Sec. 3.3　Basic Operators

exclusive OR. The XOR operator is true if one and only one of the input variables is true. It should be noted that in common language usage, XOR is normally implied. In automation applications, the OR function is by far the most widely used since it is much more readily implemented in hardware. The symbol use for OR is +, while the XOR is represented by (+).

A combination of operators again results in another useful function, NOR which is defined as the complement of the inclusive OR. In other words, the OR operation is performed and the result complemented. NOR is indicated by complementing the *entire* OR, and should be properly called OR-NOT, but the acronym NOR is widely used and accepted.

In a manner similar to that described above, the exclusive NOR (XNOR) is the complement of XOR. It is represented by complementing the *entire* XOR.

The truth tables and symbology for these four operators are shown below.

A	B	$Z = A+B$	$Z = (A+B)'$	$Z = A(+)B$	$Z = (A(+)B)'$
0	0	0	1	0	1
1	0	1	0	1	0
1	1	1	0	0	1
0	1	1	0	1	0

Figure 3.4　The Operators OR, NOR, XOR, and XNOR

3.3.4 MEMORY

The MEMORY function, while not being a basic operator in the truest sense, can be considered as a function of three variables in which the output Y depends on two inputs called SET and RESET, and the previous state, or value, of the output variable itself, called $Y_{[old]}$. No specific symbol is used to indicate the MEMORY function. Equations will be discussed later describing this function using the previously defined operators. In fact, all logic statements can be constructed using only the operators NOT, AND, and OR. The following truth table defines the MEMORY function in terms of the SET and RESET inputs.

S	R	Y
0	0	$Y_{[old]}$
1	0	1
1	1	?
0	1	0

Figure 3.5　The MEMORY Function

Note that the SET signal tends to made the output true, and RESET tends to make it false. With no input, the output is dependent on its previous state; it retains the previous state indefinitely. The MEMORY remembers its last input and retains that information until there is a new input. MEMORY is essential in remembering sequences of operation in sequential systems.

The value of MEMORY is well defined for three of the possible four input combinations. The output for the fourth condition, when both SET and RESET are activated at the same time, is dependent on the exact hardware used to implement this function, or is indeterminate. For this reason, the ? symbol is used in the truth table.

MEMORY elements can be classified into three types: type Y (or perfect), type 1, or type 0 depending on what action occurs when the SET-RESET combined input SR appears. These three types are compared below by showing the logic equations and the truth tables applicable to each. In the truth tables, the output computed from the system equation is designated Y, its complement as Y', and the complementary function computed from the equation by the application of DeMorgan's theorem as $Y'_{[c]}$. Note that Y' and $Y'_{[c]}$ are identical in all cases.

Sec. 3.3   Basic Operators   23

Type Y: $Y = SR' + Y(S'R)'$ $Y'[c] = S'R + Y'[c](SR')'$

S	R	Y	Y'	Y'[c]
0	0	Y	Y'	Y'[c]
1	0	1	0	0
1	1	Y	Y'	Y'[c]
0	1	0	1	1

Type 1: $Y = S + YR'$ $Y'[c] = S'R + S'Y'[c]$

S	R	Y	Y'	Y'[c]
0	0	Y	Y'	Y'[c]
1	0	1	0	0
1	1	1	0	0
0	1	0	1	1

Type 0: $Y = SR' + YR'$ $Y'[c] = R + Y'[c](SR')'$

S	R	Y	Y'	Y'[c]
0	0	Y	Y'	Y'[c]
1	0	1	0	0
1	1	0	1	1
0	1	0	1	1

Figure 3.6   MEMORY Types

Attention should be paid to the fact that the type Y MEMORY reduces to:

$$Y = S + YS + YR' = S + YR'$$

when expanded, and is identical to the type 1 MEMORY.

In the event that the input combination SR can never occur, that is, SET and RESET are mutually exclusive, then any type of MEMORY can be used with the same results. The type 1 MEMORY equation is usually the most desirable form since the SET and RESET signals are so readily identifiable.

3.3.5 Hierarchy of Operations

In a scalar algebra, a hierarchy of operations exists in any expression in which they are no parentheses, such as $AB + C$, where the multiplication operation is performed prior to the addition. In Boolean algebra, there is a similar hierarchy. The order in which the operations are performed is *always* NOT, then AND, and lastly OR.

For example, consider the expression $AB' + C$. When evaluating this statement using specific values, the following computations will be performed in the order stated:

$$\text{COMPLEMENT } B$$

$$\text{AND } (A \text{ and } B')$$

$$\text{OR } (C \text{ to } AB')$$

Some expressions require parentheses to denote the order of operations. In these cases, the expression contained within the parenthesis is always computed first, albeit within the hierarchy described above. In the statement $A(B + C') + D$, the order of evaluation is:

$$\text{COMPLEMENT } C$$

$$\text{OR } (B \text{ to } C')$$

$$\text{AND } (A \text{ and } (B + C'))$$

$$\text{OR } (A(B + C') \text{ to } D)$$

Notice further, in expressions containing NAND or NOR, that parentheses are implied; therefore the value of the complemented operation must be evaluated first, then complemented, then the remaining operations performed; in $(AB)' + C$, the order of evaluation is AND, NOT, OR.

3.4 POSTULATES AND THEOREMS

Essential relationships follow from the basic concept of logic:

> A statement is either true or it is false,
> it cannot be both, it cannot be neither.

These *givens* are the postulates of logic. The postulates can, in turn, be used to develop theorems which are useful in further logical manipulations.

3.4.1 Postulates of Boolean Algebra

The above descriptions of the variables that will be used in logical automation systems and the values that these variables can assume, along with the definitions of the operators discussed above, leads to a set of postulates and theorems that are recognized as an algebra. The postulates are some very basic statements concerning the nature of the algebraic system. These are drawn from the definitions of the variables and the operators. The postulates of Boolean algebra, and their *duals*, are given in Figure 3.7. The nature of the *duals* will be discussed in a subsequent section.

No.	Type	Postulate	Dual Postulate
P1	Binary Variable	$A = 0$ if $A' = 1$	$A = 1$ if $A' = 0$
P2	Complement	$1' = 0$	$0' = 1$
P3	Basic Operation	$1(1) = 1$	$0+0 = 0$
P4	Basic Operation	$0(0) = 0$	$1+1 = 1$
P5	Basic Operation	$0(1) = 1(0) = 0$	$1+0 = 0+1 = 1$

Figure 3.7 Postulates of Boolean Algebra

3.4.2 Theorems of Boolean Algebra

A set of theorems can be derived from the postulates. These theorems are the more recognizable forms of the algebra. Once a theorem has been proven or derived directly from the postulates, it can be used to derive other theorems.

There is no limit on the number of theorems that can be stated, but there are advantages to listing only those of the greatest value. Figure 3.8 shows such a list. The *duals* of the theorems are also listed.

Boolean Algebra Chap. 3

No.	Type	Theorem	Dual Theorem
		Single Variable	
T1	Complement	$A = (A')'$	$(A')' = A$
T2	Characteristic	$0(A) = A(0) = 0$	$1+A = A+1 = 1$
T3	Identity	$1(A) = A(1) = A$	$0+A = A+0 = A$
T4	Idempotent	$A(A) = A$	$A+A = A$
T5	Inclusion	$A(A') = A'(A) = 0$	$A+A' = A'+A = 1$
		Multiple Variables	
T6	Commutative	$AB = BA$	$A+B = B+A$
T7	Absorptive	$A(A+B) = A$	$A+AB = A$
T8	Reflexive	$A(A'+B) = AB$	$A+A'B = A+B$
T9	Consistency	$AB+AB' = A$	$(A+B)(A+B') = A$
T10	DeMorgan's Law	$ABC = (A'+B'+C')'$	$A+B+C = (A'B'C')'$
T11	Associative	$A(BC) = (AB)C = ABC$	$A+(B+C) = (A+B)+C = A+B+C$
T12	Distributive	$A(B+C) = AB+AC$	$A+BC = (A+B)(A+C)$
T13	Reflective	$AC+A'BC = AC+BC$	$(A+C)(A'+B+C) = (A+C)(B+C)$
T14	Transposition	$AB+A'C = (A+C)(A'+B)$	$(A+B)(A'+C) = AC+A'B$
T15	Transitive	$AB+BC+A'C = AB+A'C$	$(A+B)(B+C)(A'+C) = (A+B)(A'+C)$

Figure 3.8 Theorems of Boolean Algebra

3.5 THEOREM PROOFS

The proof of theorems is a useful exercise to gain familiarity with the theorems and to develop the techniques necessary for the algebraic reduction of expressions. There are two basic methods of proving theorems: through the use of truth tables, and by the methodical application of the postulates and previously proven theorems.

3.5.1 Proof Using Truth Tables

In the truth table method, all combinations of the independent variables are listed and values of both sides of the theorem to be proven are computed using the definitions of the operators. If both sides of the statement to be verified have *exactly* the same values for every possible combination of the inputs, then the theorem is proven. The truth table method is called *proof by perfect induction* since the listing of input combinations is exhaustive,

Sec. 3.6 Definitions

and the conclusion is drawn from a complete and perfect search. It is what is known as a *brute force* method, but it should be always be remembered that it is *perfect*, and is therefore a *benchmark* technique.

For example, consider Theorem T8: $A(A' + B) = AB$

A B	A(A'+B)	AB
0 0	0(1+0) = 0	0(0) = 0
1 0	1(0+0) = 0	1(0) = 0
1 1	1(0+1) = 1	1(1) = 1
0 1	0(1+1) = 0	0(1) = 0

Q.E.D.

Figure 3.9 Truth Table Proof (Theorem 8)

Since each value on the left side of the equation has exactly the same truth value as the corresponding one on the right, and since the search of the combinations of the input variables is perfect, Theorem 8 has been proven by perfect induction.

3.5.2 Proof Using Theorems

Theorems can also be verified by using previously proven theorems. To prove a theorem by the theorem method, one side of the equation must be made identical to the other side using only the postulates and the previously proven theorems. This method of proof requires more insight into the logical process than does the "brute force" truth table method, since several different avenues of manipulation may have to be followed on each side of the expression in order to obtain identity. There are no set guidelines or patterns for these manipulations. Experience will show which theorems are the most useful to an individual designer.

In proving theorems by this method, it is helpful to state which theorem is used to justify each step in the proof. This is demonstrated in the following examples.

Example

Consider dual theorem T7: $A + AB = A$. The left side of the expression is manipulated to make it identical to the right side, as follows:

Expression	Justification
A+AB ? A	Statement
A(1)+AB ? A	T3
A(1+B) ? A	T12
A = A	T2, T3

Q.E.D.

Figure 3.10 Theorem method proof (Theorem T=10, DeMorgen's Theorem)

Example

In some cases, it is not desirable to start with the theorem itself when proving its validity. An example of this is the following proof of Theorem T10 (DeMorgan's theorem):

Expression	Justification
$1 = A+A'$	T5
$1 = A(B+B')+A'(B+B')$	T5, T3
$1 = AB+AB'+A'B+A'B'+A'B'$	T4, T12
$1 = AB+(AB'+A'B+A'B'+A'B')$	T11
$1 = AB+A'(B+B')+B'(A+A')$	T6, T12
$1 = AB+(A'+B')$	but $AB+(AB)' = 1$ by T5, therefore
$A'+B' = (AB)'$	Substitution
	-------- Q.E.D

Figure 3.11 Theorem Method Proof (Theorem 10)

3.6 DEFINITIONS

There are certain forms in Boolean algebra that have been assigned special designations.

Variables and *complemented variables* have been previously discussed.

A *literal* is defined as the appearance of a variable or its complement in an expression; for example in the expression $AB' + C$; A, B', and C are all literals.

A *term* is two or more literals combined together by ANDs, such as $AB'C$. An *alterm* is two or more literals combined together by ORs, like $A + B' + C$. The *disjunctive* form of an expression is an ORing (or *summing*) of terms, like $AB + BC' + AB'C$. The *conjunctive* form of an expression in the ANDing (or *product*) of alterms, e.g., $(A + B)(B + C')(A + B' + C)$.

The *canonical* form is one in which all variables appear in every term or alterm. This form may either be disjunctive or conjunctive. The canonical disjunctive form may be developed through the repeated application of theorem T3 and dual T5, among others. For example, the disjunctive form shown above can be expanded as follows:

$$AB + BC' + AB'C = AB(C + C') + (A + A')BC' + AB'C$$
$$= ABC + ABC' + A'BC' + AB'C$$

In the *canonical conjunctive* form, each variable appears in every alterm. A canonical term is called a *minterm*; a canonical alterm is called a *maxterm*.

Sec. 3.7   Equation Manipulations and Reductions   29

Independent variables have no functional relationship to each other, and can form all possible truth value combinations. If the variables *A* and *B* are independent, their truth values can form the combinations 00, 01, 11, and 10 with each other.

Two variables are said to be *equivalent* if and only if (written iff) their only possible truth value combinations are 00 and 11; that is, $A = 0$ iff $B = 0$, and $A = 1$ iff $B = 1$. They are said to be *complementary* iff their truth value combinations can only be 01 or 10.

Variables are *exclusive* iff the truth value combination 11 cannot occur. An example of exclusive variables would be an actuator with limit switches at the extremes of its travel. Either one of the sensors could be ON, or they could both be OFF, but since a body cannot be in two different places simultaneously, they could never both be ON[4].

3.7 EQUATION MANIPULATIONS AND REDUCTIONS

There are several manipulations that are often necessary, even though this operation does not simplify, or reduce, the expression. These include obtaining the complement, the dual, and the *universal operator*.

3.7.1 The Complement of an Expression

The complement of an expression can be obtained by applying DeMorgan's theorem (and several others) repeatedly, and then laboriously working through the algebra; but the procedure can be significantly simplified by performing the following steps, in order:

- Complement all literals
- Interchange all 0s and 1s
- Interchange all ANDs and ORs
- Leave all parentheses as they were in the original expression

Example

Complement the following equation using the above rules

$$Z = AB'(C' + D) + EF' + (G + H)'$$

Solution:

$$Z' = (A' + B + CD')(E' + F)(G + H)$$

---

[4] Knowing this, consider the situation where simultaneous signals are received from exclusive variables in an automation system. This is called a "pathological" condition (one that cannot occur . . . but does). It means either that the variables were *not* exclusive, or that a malfunction occurred.

In the event that any confusion exists as to the separation or hierarchy of operations in the original statement, parentheses should be carefully placed to maintain its integrity. The results of such an operation on the above example would have resulted in the statement:

$$Z = (AB' \ (C' + D)) + (EF') + (G + H)'$$

Which, incidentally, could have been written using DeMorgan's theorem as:

$$Z = (AB' \ (C' + D)) + (EF') + (G'H')$$

This latter statement results in exactly the same complement as originally derived, but without treating the NOR term as a literal, and thereby avoiding some possible confusion.

3.7.2 The Dual of an Expression

The dual of an expression is only helpful in the study of postulates and theorems. There is no particular logical relationship between the value of an expression and its dual. A dual is useful since, if a theorem is derived or proven, then its dual is automatically true. The expression dual can be developed as follows:

- Interchange all 0s and 1s
- Interchange all ANDs and ORs
- Leave all parentheses as they were in the original expression

The steps required to obtain a dual are identical to those used to obtain the expression complement, except that the literals are not complemented.

3.7.3 The Universal Operators

Often one specific type of hardware is more readily available, less expensive, more compact, or for some reason, more desirable to use. In automation logic, the NAND and NOR operators, as well as one which shall be called INHIBIT and defined as equaling AB', fall into this cateogry. As a group, these are called *universal operators*.

In reviewing the postulates and theorems, it should be evident that the NOT, AND, and OR operators are necessary and sufficient to construct the entire Boolean algebra. If a MEMORY operator is added to this group, a complete control or automation system can be implemented.

While the NAND operator can be important in electronic systems, it is not readily implemented in nonelectronic automation applications; therefore only the NOR and INHIBIT functions will be considered here. The utility of these operators can be better visualized with hardware as treated in a later chapter, but the equations showing the creation of the basic operators using the two universal functions are shown in Figure 3.12.

FUNCTION	using NOR	using INHIBIT
NOT	$A' = (A+0)'$	$A' = 1(A)'$
AND	$AB = ((A+0)' + (B+0)')'$	$AB = A(1(B)')'$
OR	$A+B = ((A+B)' + 0)'$	$A+B = 1((1(B)'A')'$
MEMORY	$Y = ((S+Y)' + R)'$	$Y' = 1((1(Y)'S')'R')'$

Figure 3.12 Universal Functions

The technique for a conversion to use NOR logic elements is:

For disjunctive statements:
- Apply DeMorgan's theorem to each alterm repeatedly until all complemented literals are eliminated.
- Double negate the entire expression. The first complement forms the ultimate NOR, the second is merely a NOT operator.

For conjunctive expressions:
- Apply DeMorgan's theorem to the overall expression, then work inward using DeMorgan to eliminate the complemented literals.

Example

Convert the following expressions to NOR implementation form:

$$Z_1 = AB' + C'D'E + FG' \qquad Z_2 = (A + B')(C' + D' + E)(F + G')$$

Solutions:

$$Z_1 = (A' + B)' + (C + D + E')' + (F' + G)'$$
$$= ((A + B)' + B)' + (C + D + (C + E)' + (D + E)') + (F + (F + G)')'$$
$$= (((A + B)' + B)' + (C + D + (C + E)' + (D + E)')' + (F + (F + G)')')''$$

$$Z_2 = ((A + B')' + (C' + D' + E)' + (F + G')')'$$
$$= ((A + (A + B)')' + ((C + E)' + (D + E)' + E)'(F + (F + G)')')'$$

The technique for conversion to INHIBIT operators is not as straightforward but is equally easy, especially for the disjunctive form, which is the most widely used. Note that the OR conversion is as shown in Figure 3.12.

Example

$$Z = AB' + C'D'E + FG' = AB' + (C'E)D' + FG' = 1(1(AB')(C'E)D'(FG'))'$$

3.8 PROBLEMS

1. Write a word description for the operator ADD in scalar algebra.
2. Write a word description for the operator MULTIPLY in scalar algebra.
3. Construct the truth table for the operators AND and OR using three input variables A, B, and C.
4. Construct the truth table for the operator XOR using three input variables as above.
5. Prove Theorem T7 using a truth table only.
6. Prove Theorem T9 using a truth table only.
7. Prove Theorem T10 using a truth table only.
8. Prove Theorem T12 using a truth table only.
9. Prove Theorem T13 using a truth table only.
10. Prove Theorem T14 using a truth table only.
11. Prove Theorem T8 using the theorem method (analytically).
12. Prove Theorem T9 using the theorem method (analytically).
13. Prove the Dual of Theorem T12 using the theorem method (analytically).
14. Prove Theorem T13 using the theorem method (analytically).
15. Prove Theorem T14 using the theorem method (analytically).
16. Prove Theorem T15 using the theorem method (analytically).
17. Using the expression shown below, obtain the complement using the expression complement method, and reduce it to its disjunctive form. Check the result using the theorem method.

$$Z = AB' + BC$$

18. Obtain the complement of: $Z = C'D'(A + B') + AB' + C'D$
19. Obtain the NOR form of: $Z = A'BC + B'C$
20. Obtain the NOR form of: $Z = A'(B' + C)(C' + D)$
21. Obtain the INHIBIT form of: $Z = A'BC + B'C$
22. Obtain the NOR form of: $Z = A'B(C + D') + AD + C'D$
23. Obtain the INHIBIT form of: $Z = A'B'C + A'B'D + A'CD$
24. Convert to conjunctive form: $Z = XY' + W(X' + Z)$
25. Convert to canonical conjunctive form: $Z = A'B'C + C'D$

Chapter 4

EQUATION OPTIMIZATION METHODS

One of the goals of synthesis is to obtain an implementation of a logic expression in minimal form. After the system logic equations are derived, they must be simplified and minimized. This in turn will result in an automation system that will respond in the shortest possible time using the least amount of hardware. For this reason, it is imperative that the system designer become expert in minimization techniques. Three of these techniques will be discussed: the theorem or analytical technique, a mapping technique, and the prime implicant or tabular technique. Depending on the number of variables and the nature of the equations, one of these methods will be superior to the other two.

4.1 THE ALGEBRAIC METHOD

The techniques used to transform an original logic equation to its most reduced form by algebraic methods are very similar to those used in proving the truth of theorems. One performs the various manipulations on the independent variables until a minimal form is reached. Particular attention should be paid to the use of the word *minimal* rather than *minimum*. In order to be sure that a minimum form has been reached, it would be necessary to perform all the possible reductions on an expression, then evaluate them and eliminate all but the minimum design. This would require an exhaustive search and perfect knowledge, neither of which is practically realizable. Since this is the case, it cannot be known with any degree of certainty that any particular solution is truly "minimum".

Many factors are important when evaluating a design to determine its minimality or optimality. There are many reasons why this is a difficult objective, not the least of which is deciding exactly what is meant by *optimal*. Frequently, the cost of construction of the automation system is of prime importance. This cost depends not only on the acquisition cost of the components (including the interconnecting hardware), but also on the expense involved in the design and construction. Systems that require extensive and complicated interconnections are expensive to build, test and approve, and maintain.

Another important consideration is the reliability of the system performance. Reliability can be achieved either by using extremely reliable components or by including redundant control elements, so that the overall system will continue to perform its design mission correctly even if one or more of the individual components fail. When considering reliability, one must always remember that everything will fail eventually; the only question is when.

The third major system characteristic is the response time that is required for the automation system to react to a change in the input configuration.

Sec. 4.1   The Algebraic Method   35

If it is assumed that response time is of primary concern (within the time constraints of the machine or system being automated) and that the cost is a direct function of the number of components used, systematic synthesis procedures can be employed.

The time constraints dictate that the resulting automation system be in the form of a two-stage circuit, independent of the number of components required to provide the appropriate logical outputs. Algebraically, this means that the logic equations must be written in *either* disjunctive or conjunctive form. The main reason for insisting upon the minimum response criterion is that it enables using a formal synthesis scheme by severely restricting the type of system that can be considered in the search for an optimum. For any given automation system, there are many different two-stage circuits that can be constructed. Usually, the system requiring the fewest components is the one that is desired. This is obtained by associating a cost function with each disjunctive or conjunctive expression corresponding to the design function.

There have been many attempts to define this cost function. If the system is to be assembled using discrete components with two or fewer inputs, such as valves, relays, or diodes, the relative value of the circuit can be evaluated by summing the number of appearances (literals) and the number of operators. This cost can be expressed directly in terms of the control equation.

Example

Determine the minimal form of the equation

$$Z = A + ABCD + A'BCD + B'C'D'$$

Solution: By applying Theorem T9 to the second and third terms, the statement becomes

$$Z = A + (A + A')BCD + B'C'D' = A + BCD + B'C'D'$$

(4 literals + 9 operators, score = 13)

$$= A + BCD + (B + C + D)'$$ (4 literals + 7 operators, score = 11)

$$= A + (B'C)D' + B'C'D'$$ (5 literals + 6 operators, score = 11)

Note that the above exercise not only demonstrates the scoring technique (the last statement uses both C and C' as literals, and the INHIBIT operator), it also shows the use of algebraic simplification techniques. Other reductions could also have been made, but it is not the purpose of this exercise to perform an exhaustive search, only to obtain a reasonable solution.

As previously stated, the algebraic technique involves applying the postulates and theorems, as well as the definitions of the operators, to the automation equation. While there are no set rules as to how to do this (experience is the best guide), the scheme of expanding a statement to disjunctive form, and systematically grouping and reducing to obtain a minimal expression usually works. In the event that electronic controls are available, an expression in conjunctive form for implementation with NAND logic can be derived.

Examples:

Reduce the following expressions to their minimal form.

$$Z_1 = AB + ACD + AC' + AD' + E$$

$$Z_2 = (A + B + C)(A + B' + C')(A + B' + C)$$

$$Z_3 = BD + AD' + AB'CD$$

Solutions:

$$Z_1 = A(B + CD + C' + D') + E$$
$$= A(B + (C' + CD) + D') + E = A(B + C' + D + D') + E$$
$$= A(B + C' + 1) + E = A(1) + E = A + E$$

(2 literals + 1 operator, score = 3)

$$Z_2 = (A + B + C)(A + AB' + AC + B' + B'C + AC' + B'C')$$
$$= (A + B + C)(A(1 + \cdots) + B'(1 + \cdots))$$
$$= (A + B + C)(A + B') = A + AB' + AB + AC + B'C$$
$$= A(1 + \cdots) + B'C = A + B'C$$

(3 literals + 2 operators, score = 5)

$$Z_3 = AD' + D(B + AB'C) = AD' + D(B + AC) = A(D' + CD) + BD$$
$$= AD' + AC + BD$$

(4 literals + 5 operators, score = 9)

$$Z_3 = ((AD')'(AC)'(BD)')'$$ (5 literals + 4 operators, score = 9)

4.2 MAPPING METHOD

Boolean algebra shows the relationships among the two-valued variables involved in a logical expression. Since the relationships become much less apparent as the logical algebraic expressions become more complex, a more descriptive technique would be advantageous. One of the methods that has

Sec. 4.2 Mapping Method

proven to be desirable is the mapping method developed by J. Venn and M. Karnaugh. This method has the advantages of being flexible, fast, and easily learned.

Maps are a form of truth table where each cell in the map corresponds to each possible canonical term in the truth table. These cells are arranged into a matrix which represents the universe. Each cell occupies a unique position which is determined by the input combination that it represents. The arrangement of cells is such that simplified logic expressions can be readily read directly from the map.

Prior to the discussion of the arrangement and construction of maps, it is appropriate to mention the naming or coding of cells. The type of code that will be used is called a *Gray* code.

A Gray code is a binary code that is of significant value in automation synthesis. In a Gray code, the progression of numbers is arranged such that the $(n + 1)^{th}$ number differs from the nth number by a change in one and only one digit. The order or location of the digits has no particular significance, except in logic synthesis; each digit represents a variable. A Gray code is not a counting code.

A Gray code is not unique; any code that enables a system to move from one state (input combination) to another by a change in one and only one variable is a Gray, or adjacent code. Note that lines 0 and 15 in Figure 4.1 are adjacent. One can move from either line to the other through the change of only one variable.

In a *left* Gray code, the most rapidly changing, or least significant, variable is on the left. In a "right" Gray code, the least significant variable

Line Number	Gray Code (Left) A B C D	Gray Code (Right) A B C D
0	0 0 0 0	0 0 0 0
1	1 0 0 0	0 0 0 1
2	1 1 0 0	0 0 1 1
3	0 1 0 0	0 0 1 0
4	0 1 1 0	0 1 1 0
5	1 1 1 0	0 1 1 1
6	1 0 1 0	0 1 0 1
7	0 0 1 0	0 1 0 0
8	0 0 1 1	1 1 0 0
9	1 0 1 1	1 1 0 1
10	1 1 1 1	1 1 1 1
11	0 1 1 1	1 1 1 0
12	0 1 0 1	1 0 1 0
13	1 1 0 1	1 0 1 1
14	1 0 0 1	1 0 0 1
15	0 0 0 1	1 0 0 0

Figure 4.1 Gray Codes

is on the right. A Gray code could also have one of its inner variables changing most rapidly. The left Gray code is used throughout this work, even though the right Gray is most often seen in other sources. The right Gray is more compatible with binary counting codes, while the left Gray is more compatible with computer codes used in logical synthesis. In addition, the left Gray is much more easily distinguished from counting codes, which will be discussed later.

In physical systems, it is extremely unlikely that two or more variables would change states at exactly the same instant (simultaneously). Consequently, the probability of a single event change at any instant is very high. Therefore, it is felt that a Gray code models the real world much more satisfactorily than any other coding.

A simple way to remember the development of a Gray code is to notice, as is shown in Figure 4.1, that the least significant column starts with a 0 followed by two 1s, then two 0s, and so on. The next most significant column starts with two 0s, followed by four 1s, followed by four 0s, and so on. Each succeeding column is similarly patterned until the coding is complete. Further, it should be noted that for n variables, there are exactly 2^n combinations of variables, or states.

At this point, one more symbol of a binary variable is introduced, namely the *optional* value. Often the value of a variable does not affect the logic of the system, either in that the combination can never physically occur or, if it does occur, one is indifferent to the result. This is called a *don't care* or *optional* state, and is indicated by a dash "-". The dash will be used in several different contexts; these will be discussed in other sections. In all cases, however, it refers to the optional state. The dash used in a logic state, such as 01–1 represents the state $A'B(C + C')D$ or $A'BD$. In this case the variable C does not appear in the term since we *don't care* about it.

As stated before, algebraic expressions that contain n variables can be represented by maps containing 2^n cells. The conventional arrangement of cells is $2^a \times 2^b$ where $a + b = n$. For example, a three variable map would contain 2^3 or 8 cells arranged in a $2^1 \times 2^2$ configuration while a five variable map would contain 2^5 or 32 cells arranged in a $2^2 \times 2^3$ configuration, and so forth. The identification of each cell is accomplished by using a left Gray code with each position corresponding to a variable. Examples of map layouts and cell identification are shown in Figure 4.2.

The ability to quickly and accurately simplify a Boolean expression through the use of a Karnaugh map is a function of the unique arrangement of the cells in the map. The criterion of cell assignment is that two cells that are next to each other on a map can differ from each other by one and only one state change in any one variable. This property is called *adjacency*. It should be noted that each cell in a map is adjacent to every adjoining cell. Cells at the ends of rows or columns are adjacent to not only the adjoining,

Sec. 4.2   Mapping Method

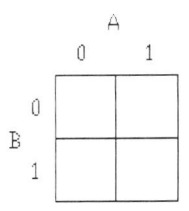

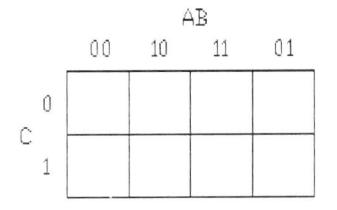

Figure 4.2a   Two Variable Map

Figure 4.2b   Three Variable Map

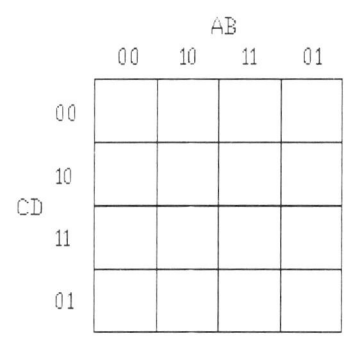

Figure 4.2c   Four Variable Map

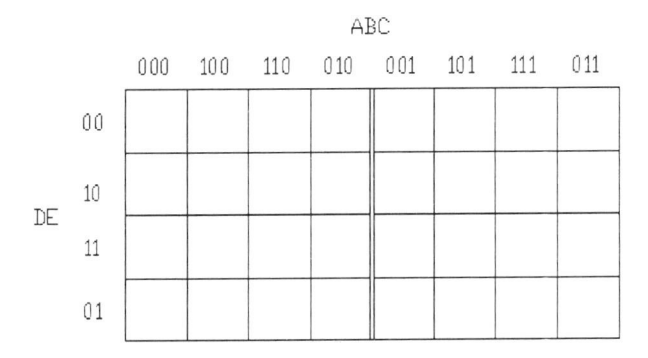

Figure 4.2d   Five Variable Map

but also to the cells at the opposite ends of the rows and/or columns (see Figure 4.2c).

These maps also can be three-dimensional. Each plane contains no more than a 4 by 4 map, and all subsequent maps are stacked above or below that plane to form the third dimension of adjacency. Figure 4.2d illustrates this, in that the right-hand 4 by 4 map should be visualized as being above or below the 4 by 4 map on the left-hand side. Note also that this adjacency requirement imposes a practical limitation on the number of independent variables that can reasonably be mapped. The maximum usable map consists of a cube 4 by 4 by 4 containing six independent variables.

The use of a map starts with the entering of information into the cells. In this process, the map acts as a truth table, albeit in a highly compressed form. Each cell depicts either a canonical product or minterm, in which case the entire map shows a disjunctive expression; or each cell represents a canonical sum, or maxterm, in which case the map shows a conjunctive expression. For the purposes of this discussion, the disjunctive form will be described.

All cells representing the minterms that are included in the true function are identified by placing a 1 in the appropriate cell. The cells representing the false function are usually left blank, although it is understood that a 0 really occupies that location.

Once the required information has been entered, the simplified function can easily be read directly from the map. The technique is to group the cells containing 1s (or 0s if the complementary function is desired) into groups of $2^m \times 2^n$, where m and n are always integers. If two adjacent cells can be grouped then one variable can be removed from the minterm; if four adjacent cells can be grouped, another variable can be removed. The general rule is that one variable can be removed from a minterm for each increase of 1 in either "m" or "n", that is $BD = BD(A + A')(C + C')$.

Example 4.2a

$Z = BD + AD' + AB'CD$

$= BD(A + A')(C + C') + AD'(B + B')(C + C') + AB'CD$

$= ABCD + ABC'D + A'BCD + A'BC'D + ABCD' + ABC'D' + AB'C'D'$
$+ AB'CD$

$Z = AD' + AC + BD$

Example 4.2a

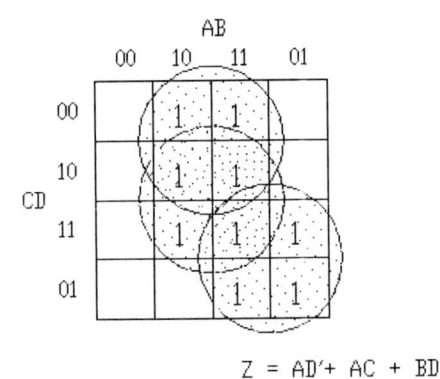

Sec. 4.2   Mapping Method   41

Example 4.2b

$$Z = (A + B + C)(A + B' + C')(A + B' + C)$$

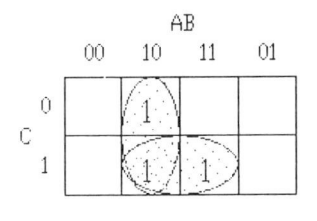

```
         AB
      00  10  11  01
   0  [ ][1][ ][ ]
C
   1  [ ][1][1][ ]
```

$Z = (A+B')(A+C) = A + B'C$

Example 4.2b

Example 4.2c

$$Z = AB + ACD + AC' + AD' + E$$

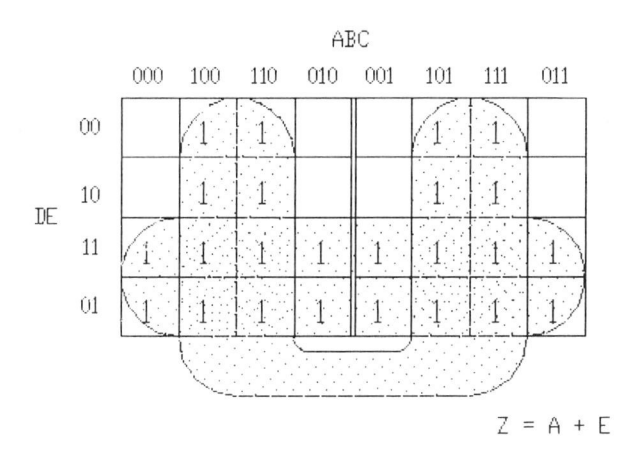

```
              ABC
      000 100 110 010 001 101 111 011
   00  [ ][1][1][ ][ ][1][1][ ]
   10  [ ][1][1][ ][ ][1][1][ ]
DE 
   11  [1][1][1][1][1][1][1][1]
   01  [1][1][1][1][1][1][1][1]
```

$Z = A + E$

Example 4.2c

Example 4.2a-c illustrates a series of maps from which simplified expressions can be read. Note that while examples *a* and *c* treat disjunctive expressions, example *b* WAS expressed conjunctively. This latter example was expanded into disjunctive canonical form for purposes of illustration. It is left to the reader to compare this to the other examples.

The final product of the map extraction was then reduced to the disjunctive form. This was done since it seemed apparent that this simplification should be made. It should be realized that the appropriate technique should always be used. Quite commonly, an expression can be readily partially reduced algebraically, then mapped, and then converted algebraically to a more optimal form.

The rule for the extraction of a minimal statement from a map can be stated as:

What goes in, must come out.

That is, if a disjunctive statement is mapped, then a disjunctive statement must be extracted; if a conjunctive statement is mapped, then a conjunctive statement must be extracted. Note also that visualization of minimal groups can often be clarified by *circling* the appropriate cells.

The process of grouping continues until all the suitable cells have been included *at least once* in a term or alterm. Some cells may not fit into any group. In this case, the minterm or maxterm must be used. Some terms may fit into a number of groups which is quite acceptable as was stated in Postulate 4 and Theorem 4.

In the preceding discussions, the assumption was made that it is critical to the operation of an automation system that the states of all variables be defined. As was stated earlier, there are cases when there is no functional significance to the state of a variable or combinations of variables; that is they are *optional*. To repeat, this condition exists if a particular combination can never physically occur or if the function of the system being controlled is truly indifferent to the input state.

A system that illustrates the use of the optional states is shown in Example 4.2d. Consider the condition where an automation system output can be defined by the statement:

$$Z = A'C + ABCD$$

Furthermore, it is known that the input combinations $ABC'D$, $ABCD'$, $AB'CD$, and $AB'CD'$ can never physically occur. Since the system designer is indifferent to these four input states, they can be assigned *any* value with-

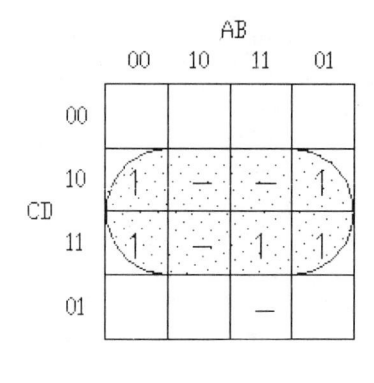

$$Z = C$$

Example 4.2d

out affecting the system's operation. Upon examination of the map, it can readily be seen that if three of the four optional states were assigned a value of 1, and the remaining state assigned a 0 by default, the simplified result would be $Z = C$.

The above technique demonstrates one of the extremely powerful features of the mapping technique. It allows the designer to visualize the relationships among the functions and the optional states, thereby enabling simplifications that otherwise might not be apparent.

4.3 SIMPLIFICATION USING PRIME IMPLICANTS

Any control equation can be written in disjunctive canonical form by summing all the minterms. This procedure would normally not produce an optimal statement since some of the literals in the minterms are redundant and could be deleted through various simplification procedures. The efficiency of the design procedure using prime implicants depends on the fact that only certain terms need to be considered. It can be demonstrated that a minimum cost system always has the property that, if one of the literals in any term is removed, the system will no longer provide the correct output function. When the number of operators is to be minimized, it can also be shown that an optimal system can be obtained which also has the property that the removal of one of the inputs results in an incorrect output.

4.3.1 Definitions and Proofs

A function $f(x_1, x_2, \ldots, x_n)$ is said to *include* another function $g(x_1, x_2, \ldots, x_n)$ if, for any assignment of values to $x_1, x_2, \ldots, x_n$ which causes g to equal 1, f will also equal 1.

A prime implicant of a function f is a product of literals having the following properties:

- The function f includes this product of literals, and
- If any one of the literals is removed from the product, the function f does not include the product of the remaining literals.

Prime Implicant Theorem. This theorem states that if the cost of a two-stage system is defined such that removing a literal from the corresponding algebraic expression always decreases the cost, then the algebraic expression corresponding to a minimal-cost two-stage system will always be the sum of prime implicants or its derivative conjunctive expression.

Proof. Consider that there exists a minimal-cost system for which the corresponding disjunctive algebraic expression is not a sum of prime implicants. Then at least one of the terms must not be a prime implicant. This means that it must be possible to remove one of the literals from this term and have the product of the remaining literals included in this function. The original term which was not a prime implicant can therefore be replaced by a term involving one less literal. Thus, a new disjunctive expression is formed which contains that same number of terms, but one less literal. Since this expression still expresses the same function but has a lower cost than the original expression, the original expression cannot have been a minimal sum.

Generalized Prime Implicant Theorem. When a definition of system cost is used such that the cost does not increase when a literal is removed from a term, there is at least one minimal system for which the corresponding algebraic expression is a sum of prime implicants.

Proof. Consider that there exists a minimal system for which the corresponding disjunctive algebraic expression is not a sum of prime implicants. Then there must be at least one term that is not a prime implicant. This means that at least one literal in this term can be deleted and the remaining term will still be included in the original function. The term with one literal removed can then be substituted for the original term in the disjunctive expression. A new expression results which has the same number of terms, but one fewer literal than the original expression. Since the cost does not increase when a literal is removed, the cost of the system corresponding to the new expression is at least as small as the system corresponding to the original expression. If this procedure is repeated until a disjunctive expression of prime implicants is attained, then a minimal-cost system is derived.

These theorems merely state that a minimal-cost system for either definition of cost discussed above can always be obtained when it is not possible to remove any literal without changing the function realized by the system. This discussion shows that the first step in obtaining a minimal-cost system must be concerned with finding the prime implicants of the function for which the system is to be realized.

4.3.2 Coding

In section 4.2, it was pointed out that the Gray code is used where adjacency is important, such as in moving from one state to another through the change of only one variable. In the subsequent prime implicant analysis, it is important to be able to *name* the states. To do this, another binary code will be used.

The most obvious binary code in a numbering or naming system is the counting code that arises from a *base two* (base$_2$) system. Since this is a counting code, it is discrete and only deals with integers. Each location of the symbol has a particular exponential significance to the base 2. In other

Sec. 4.3   Simplification Using Prime Implicants   45

Decimal base₁₀	Binary base₂ $2^3\ 2^2\ 2^1\ 2^0$
0	0 0 0 0
1	0 0 0 1
2	0 0 1 0
3	0 0 1 1
4	0 1 0 0
5	0 1 0 1
6	0 1 1 0
7	0 1 1 1
8	1 0 0 0
9	1 0 0 1
10	1 0 1 0
11	1 0 1 1
12	1 1 0 0
13	1 1 0 1
14	1 1 1 0
15	1 1 1 1

Figure 4.3 Decimal to Binary Conversion

words, the *rightmost* digit represents 2^0. This digit, and all others, may be either 0 or 1. The next digit to the left represents 2^1, the next 2^2, and so on to 2^n. A decimal conversion to binary coding is shown in Figure 4.3.

4.3.3 Derivation of the Prime Implicants

There are several usable schemes for determining the prime implicants of any function, one of which is the Quine-McClusky method[25] which will be described. This is not necessarily the most efficient method, but it does demonstrate the feasibility of obtaining the prime implicants and also the practicality of the use of the micro-computer in automation design. The system illustrated in Figure 4.4 will be continually referenced in this discussion. The disjunctive expression which corresponds to a minimal-cost system will be called a "minimal sum". There are several disjunctive expressions that could result in a minimal cost system, each of which is a minimal sum.

Consider the first term of the minimal sum shown in Figure 4.4, *abce*. Note the variable *d* is not present; that is if *a*, *b*, *c*, and *e* are all equal to 1, then the value of the term will equal 1 regardless of the value of *d*. Similarly, in the fourth term *a'b'c'* both *d* and *e* are missing, and if *a*, *b*, and *c*

Figure 4.4a

Set	a	b	c	d	e	
0	0	0	0	0	0	✓
1	0	0	0	0	1	✓
2	0	0	0	1	0	✓
3	0	0	0	1	1	✓
7	0	0	1	1	1	✓
14	0	1	1	1	0	✓
22	1	0	1	1	0	✓
15	0	1	1	1	1	✓
23	1	0	1	1	1	✓
29	1	1	1	0	1	✓
31	1	1	1	1	1	✓

Figure 4.4b

Sets	a	b	c	d	e	
0 1	0	0	0	0	–	✓
0 2	0	0	0	–	0	✓
1 3	0	0	0	–	1	✓
2 3	0	0	0	1	–	✓
3 7	0	0	–	1	1	
7 15	0	–	1	1	1	✓
7 23	–	0	1	1	1	✓
14 15	0	1	1	1	–	
22 23	1	0	1	1	–	
15 31	–	1	1	1	1	✓
23 31	1	–	1	1	1	✓
29 31	1	1	1	–	1	

Figure 4.4c

Sets	a	b	c	d	e
0 1 2 3	0	0	0	–	–
7 15 23 31	–	–	1	1	1

Figure 4.4c

Set Name		0	1	2	3	7	14	22	15	23	29	31
*	0, 1, 2, 3	[x]	[x]	[x]	(x)							
**	7, 15, 23, 31					x			(x)	(x)		(x)
*	29, 31										[x]	(x)
*	22, 23							[x]		(x)		
*	14, 15						[x]		(x)			
	3, 7				(x)	x						

Figure 4.4d (Prime Implicant Table)

Figure 4.4 Determination of Optimal Form for Z = a'b'c'd'e' + a'b'c'd'e + a'b'c'de' + a'b'c'de + a'b'cde + a'bcde' + b'cde' + a'bcde + ab'cde + abcd'e + abcde + abce + ab'cd + a'bcd + a'b'c' + cde

are all equal to 0, then the value of the term is 1 regardless of the values of *d* and *e*.

One method of determining the prime implicants consists of first finding which minterms can have one of their literals removed without changing the value of the term. Then the terms with one variable missing are considered to determine which of these can have another literal removed; then the terms with two variables missing, and so on. The terms from which it is no longer

Sec. 4.3 Simplification Using Prime Implicants

possible to remove a variable without changing the value of the expression are the prime implicants.

A Boolean function is commonly fully specified by means of a truth table. It is also possible to specify a function by listing only those rows of the truth table whose function value is 1 or is 0, or to whose columns one is not indifferent (see Figure 4.4a). Note that all the terms in this table are minterms; no optional values are permitted. A variable which corresponds to a 0 in a row is a complemented variable, one which is represented by a 1 is uncomplemented.

If two minterms are identical in all but one variable, these minterms can be combined into a new term through the application of Theorem T9.

The first step in the procedure is to prepare the truth table in a specific order by binary number equivalents as in Figure 4.4a. Note that the rows are grouped by the number of 1s appearing in each row. This facilitates the search procedure since two rows which agree in all but one variable can differ by only one in the number of 1s contained. It is therefore only necessary to compare each row with all the rows containing only one more 1 than itself.

The next step in the procedure is to compare the rows in the table as described above to determine whether any pairs agree in all but one variable. If such a condition is discovered, a new row is created in which the variable of disagreement is designated by the optional dash, as in Figure 4.4b. Whenever two rows are found to combine, a check mark should be placed next to each row to designate that neither is a prime implicant.

The process of comparing pairs of rows is continued until all pairs which can be combined are combined. Any rows which remain unchecked are prime implicants. A table such as is shown in Figure 4.4b results from this process. Each row contains one and only one optional entry. This table is a listing of the terms that are included in the original function, with one missing variable.

Each pair of rows in Figure 4.4b must be analyzed for the removal of another variable. Two rows which contain a single dash each can be combined only if the dashes in both rows occur for the same variable and the other entries of the two rows are identical except for one variable. When these criteria are satisfied, a new row is formed having two dashes; one corresponding to the original optional variable, the second to the variable of disagreement. For example, rows for sets 0/1 and 2/3 of Figure 4.4b can be combined to form row 0/1/2/3 of Figure 4.4c. It should also be noted that the same result would have been obtained by combining rows 0/2 and 1/3 of Figure 4.4b. All the rows that are used to construct new rows are marked with check marks as above, since it is known that these are not prime implicants. When all rows with one optional variable are compared and noted, a new table comparable to that in Figure 4.4c will have been formed.

It now becomes necessary to compare all the rows formed in the new table. Each of these rows contain two optional terms, and the criteria for combination are comparable to those stated above. This process is continued until no new rows can be formed. The rows that remain unmarked are the prime implicants.

It is only necessary to include those prime implicants that are not redundant in the final statement. Because of this, a scheme is required that enables the selection of the prime implicants to be used.

4.3.4 Forming the Optimal Expression

The prime implicants which are included in the minimum sum expression are selected by means of a prime implicant table. Each column of this table corresponds to one of the minterms in the *original* statement. Each row corresponds to one of the prime implicants, and is arbitrarily labeled as in Figure 4.4d. An x is placed at the intersection of a row and column only if the prime implicant corresponding to the row includes the term corresponding to the column. Figure 4.4d illustrates a prime implicant table for the function of Figure 4.4a. Enough prime implicants must be included in the minimal sum so that each term of the disjunctive canonical expression is included in at least one of the chosen prime implicants. This means that a set of rows must be chosen so that each column has at least one x in one of the chosen rows. If any column contains only a single x, the corresponding row must be included in the minimal sum. Rows containing the only x for any column are called *essential rows* or *essential prime implicants*. In Figure 4.4d, the first, third, fourth, and fifth rows are the essential prime implicants.

The procedure to follow in this determination is as follows.

- Mark each x that appears once in a column with square brackets [].
- Mark any x that appears in a row containing a square bracketed x with parentheses ().
- Mark all unbracketed xs that appear in columns containing bracketed or parenthesized xs with braces {}.
- Mark all rows containing bracketed xs with an asterisk *; these are the minimum essential rows.
- Consider only those columns containing unbracketed xs. Select appropriate rows containing these unbracketed xs for inclusion in the final statement, and double asterisk **.

For example, consider Figure 4.4d. The xs contained in columns 0, 1, 2, 3, 14, 22, and 29 are square bracketed since they appear in each column only once. The unbracketed xs in the first, third, fourth, and fifth rows are then parenthesized. Finally, the unbracketed xs in columns 3, 15, 23, and

Sec. 4.3 Simplification Using Prime Implicants   49

31 are put within braces, leaving only the xs in column 7 unmarked. This indicates that either the second or sixth row is optionally essential. The selection is then made, based on the number of literals appearing in the row. The sixth row generated from minterms 3/7 (from Figure 4.4b) contains four literals, $a'b'de$; while the second row, consisting of minterms 7/15/23/31, contains only three, cde. The second row is therefore the optimal choice, and is double asterisked.

In practice, the function for which a system is to be designed is frequently incompletely specified. Some rows of the truth table are optional, as described in section 4.2 and illustrated in Example 4.2; that is, a system which produces either a 0 or a 1 for an unspecified row is acceptable. In order to take this into account when developing a functional relationship, the unspecified rows are included with the specified rows when obtaining the prime implicants. When the prime implicant table is formed however, only the specified rows are included. This procedure will result in a minimal expression corresponding to the original incompletely specified function.

4.3.5 Functions Expressed Non-Canonically

Quite often a control function is expressed in a form other than the disjunctive canonical. In order to implement the scheme shown above to obtain a minimal expression, it is essential that the logic function ultimately be stated in truth table form. There are several methods by which this can be accomplished. A noncanonical expression can always be expanded to disjunctive canonical form through the use of the theorems. A computer program named TRUTAB, written in BASIC for use with IBM/PC compatible microcomputers, is included in the appendices. This program will enable the user to input a logical specification, and will produce the most efficient output statement. Examples of the application of this program are given in a later chapter.

4.3.6 Multiple Output Systems

If the system requirements dictate that more than one output is required, it is frequently possible to combine or share terms among the several system outputs. In a disjunctive expression it is often possible to have the identical term used repeatedly in the different output statements. Consider the following problem:

$$Z_1 = a'b'c + ab'c' + ab'c + abc$$
$$Z_2 = a'b'c + a'bc' + a'bc + abc$$
$$Z_3 = a'bc' + abc' + abc$$

	a	b	c
1	0	0	1 ✓
4	1	0	0 ✓
5	1	0	1 ✓
7	1	1	1 ✓

Z_1

	a	b	c
1	0	0	1 ✓
2	0	1	0 ✓
3	0	1	1 ✓
7	1	1	1 ✓

Z_2

	a	b	c
2	0	1	0 ✓
6	1	1	0 ✓
7	1	1	1 ✓

Z_3

	a	b	c
1 5	–	0	1
4 5	1	0	–
5 7	1	–	1

Z_1

	a	b	c
1 3	0	–	1
2 3	0	1	–
3 7	–	1	1

Z_2

	a	b	c
2 6	–	1	0
6 7	1	1	–

Z_3

	a	b	c
1	0	0	1
7	1	1	1

$Z_1 Z_2$

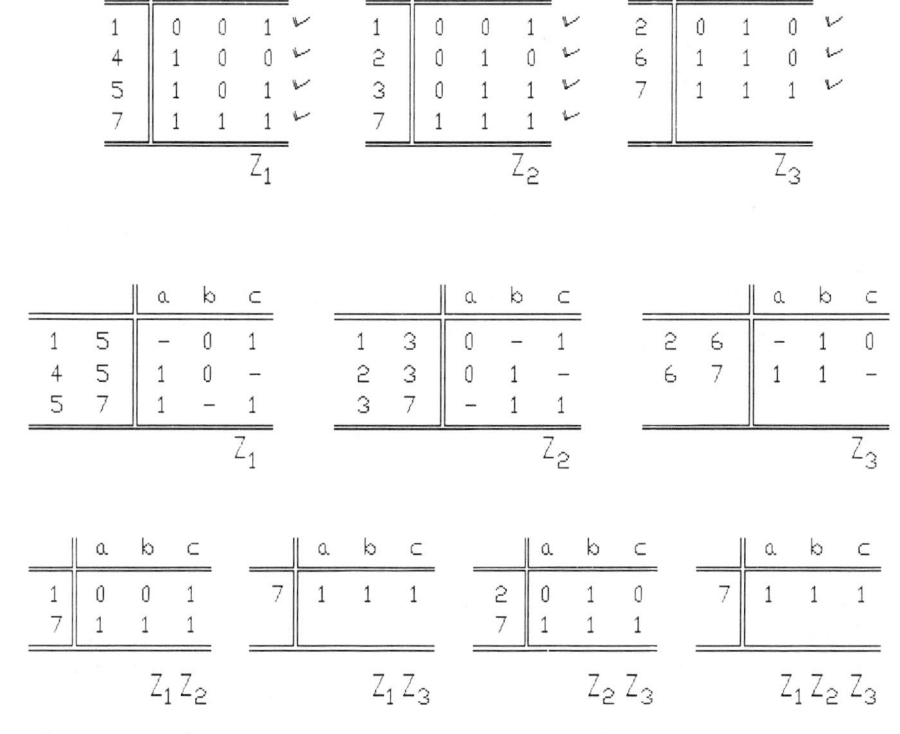

	a	b	c
7	1	1	1

$Z_1 Z_3$

	a	b	c
2	0	1	0
7	1	1	1

$Z_2 Z_3$

	a	b	c
7	1	1	1

$Z_1 Z_2 Z_3$

Figure 4.5 Prime Implicant Determination

In order to design a minimal-cost multiple output system, it is necessary to form all the prime implicants for each of the individual output functions, but this alone is insufficient. It is also necessary to determine the prime implicants required for any and all combinations of the outputs, as is shown in Figure 4.5.

A prime implicant table is then constructed in which each column corresponds to one of the minterms for the disjunctive statement for one of the outputs. If one of the minterms appears in more than one of the outputs there will be more than one column in the prime implicant table corresponding to this minterm. For example, in Figure 4.6, the minterm corresponding to the first row of the truth table appears in both Z_1 and Z_2. Therefore, this column appears under both of the outputs.

Each row of the table corresponds to a prime implicant of one or a combination of several of the outputs. The minimal-sum expressions are developed from this table by using the same rules outlined in section 4.3.3. Figure 4.6 shows the expressions that result when this scheme is applied to the functions stated in Figure 4.5.

Sec. 4.4 Summary

		Z_1				Z_2				Z_3			
		1	4	5	7	1	2	3	7	2	6	7	
	A = b'c B = ab' C = ac	× [×] 	⟨×⟩ ⟨×⟩ ⟨×⟩	 ×									Z_1
*													
	D = a'c E = a'b F = bc					× [×] 	⟨×⟩ ⟨×⟩ ⟨×⟩	 ×					Z_2
*													
*	G = bc' H = ab									[×] 	⟨×⟩ ⟨×⟩	 ×	Z_3
**	I = a'b'c J = abc	[×] 		 ×		[×] 			 ×				$Z_1 Z_2$
	K = abc		×									×	$Z_1 Z_3$
	L = a'bc' M = abc						× 		 ×	× 		 ×	$Z_2 Z_3$
**	N = abc		[×]				[×]				[×]		$Z_1 Z_2 Z_3$

Z_1 = ab' + a'b'c + abc Z_2 = a'b + a'b'c + abc Z_3 = bc' + abc

Figure 4.6 Prime Implicant Table (data from Figure 4.5)

4.4 SUMMARY

It has been demonstrated that there are three principal schemes by which minimal expressions can be derived from statements of automation problems: algebraic, mapping, and prime implicant techniques. All of these work but, depending on the number of inputs and outputs and the complexity of the problem, one will be more desirable than the others.

The algebraic technique is fast and reliable, but is cumbersome and inconvenient for systems with many input variables, or for multiple output systems.

If the system to be designed has six or fewer input variables, the mapping techniques are the most desirable due to their flexibility, simplicity, and speed of use.

The prime implicant method is the most universally applicable technique for the optimization of combinational automation systems. It has few if any limitations, but its implementation by hand is time consuming and awkward. In the event that it is desirable to use this technique, it can most easily be implemented through the use of a computer and the appropriate software. A program called PRIMIMP is listed in the appendices. This program will perform a prime implicant analysis on a single output system. Examples of its operation are shown in a later chapter.

The decision on which design technique to use is based on the above criteria, the availability of computational aids, and the personal preferences of the designer.

4.5 PROBLEMS

Using algebraic techniques *only*, simplify the following expressions:

1. $Z_1 = a + a'b + (a + b)'c + (a + b + c)'d$
2. $Z_2 = ab' + ac + bcd + d'$
3. $Z_3 = a' + a'b' + bcd' + bd'$
4. $Z_4 = ab'c + (b' + c')(b' + d') + (a + c + d)'$

Using *only* mapping techniques, simplify the following:

5. Z_1 in problem 1 above
6. Z_2 in problem 2 above
7. Z_3 in problem 3 above
8. Z_4 in problem 4 above

Simplify the following expressions using *only* Karnaugh maps:

9. $U = ab'(c + d') + b'c'd + a'b'c'd' + abc + a'bc'd$
10. $V = acd + abc'd' + ac'd + ab'c + abc'd$
11. $W = a'bc' + ad + bc'd'$
12. $X = (a + b + c)(a' + b + d)(b + c' + d')(a' + c + d')$
13. $Y = a'b'c'd + abc'e + b'cd'e' + a'bc'd + a'bc'e + b'cd'e$
14. $Z = ab'f' + cde + abcd'e + ab'ef + a'e'f + ac'e'f$

Using prime implicant techniques *only*, simplify the following:

15. U in problem 9 above
16. V in problem 10 above
17. W in problem 11 above
18. X in problem 12 above
19. Y in problem 13 above
20. Z in problem 14 above

Sec. 4.5   Problems

21. $Z_A = b'c' + abd' + ab'c + b'cd + bcd' + a'b'c$
22. $Z_B = a'b'd + abc' + bd + bcd'$
23. $Z_C = a'b'd' + abc' + ab'cde' + ab'ce + a'bcde' + a'bcd'e + a'bcde$
24. $Z_D = (b + d')(c + d)(a + b')$

Obtain the optimal form of the following system specifications using Karnaugh maps *only*. Make appropriate use of the "don't care" terms. If an OFF statement is given, assume that the unspecified minterms are optional.

25. $Z = ab'c + bcd' + abd \qquad Z_{\text{off}} = ab'c' + a'c'd'$
26. $Z = AB'F' + CDE + ABCD'E + AB'EF + A'E'F + AC'E'F$
    $Z_{\text{d.c.}} = ABC'F' + ABCE'F' + ABC'EF$

Same instructions as in problems 25 and 26, except use only the prime implicant method.

27. $Z = ab'f' + cde + abcd'e + ab'ef + a'e'f + ac'e'f$
    $Z_{\text{off}} = a'b'c'e'f' + a'cd'ef' + a'c'd'ef$
28. $Z = (B'+C'+E+F)(A+B'+C'+E')(A'+B+C'+F)(A+C'+D+E)$
    $Z_{\text{off}} = A'CD'E' + A'B'CDEF'$

Chapter 5
IMPLEMENTATION OF LOGIC FUNCTIONS

There are two constraints that are essential when designing a control or automation system.

- An optimal system must do exactly what is required, no more, no less; as well and as reliably as is necessary; at the lowest possible cost.
- A logic system has no internal preference regarding its implementation.

These two constraints recognize that there are many different hardware systems that can be used to implement control or automation. The choice is more likely to be a function of the nature of the input and output signals, the service requirements and environmental limitations, and the overall cost than the unique capabilities and niceties of any particular hardware type.

The numbers and types of components used to sense physical parameters and perform logical operations have expanded so rapidly that any attempt to list specific items is both futile and inappropriate. Readers who are concerned with sensing devices are referred to various manufacturers' literature and the many publications of the *Instrument Society of America.* For those interested in logic control devices, various sources are available. It is suggested that publications of the *Institute of Electrical and Electronic Engineers* and the *Fluid Power Institute* and trade journals such as *Control Engineering* and *Hydraulics and Pneumatics* be consulted. In addition, various works on the nature and application of various hardware systems are referenced in the bibliography.

For the purposes of this text, it is enough to say that devices capable of sensing the presence or absence of almost any physical property are readily available.

Logic control and automation devices will be described by function and general operation rather than by proprietary types. While the operating characteristics of the various hardware systems will be considered, it is felt that the choice of a specific component or supplier is best left to the designer.

5.1 LOGIC IMPLEMENTATION

It has already been demonstrated that a complete logic system can be developed using only AND, OR, NOT, and MEMORY; or devices that will implement one of the universal operators NAND, NOR, or INHIBIT; or some combination of these. It is therefore necessary to define individual pieces of hardware in terms of the logic operators.

The devices that will be discussed will be categorized by the nature of the logical inputs and outputs and the requirements for external sources of power. Inputs and outputs to be considered will be mechanical, electrical, fluid power, fluidic, and electronic. Devices that require an external source of power are called active; those that draw the required power from the

Sec. 5.2   Mechanical Input, Mechanical Output Systems   57

logical inputs are called passive. It should be noted that all devices that perform a NOT as part of their operation are active, since an output signal cannot be physically created from nothing.

5.2 MECHANICAL INPUT, MECHANICAL OUTPUT SYSTEMS

In the event that a physical displacement is the required input to an automation system, (the end-of-travel of a machine tool table), mechanical sensors and logic are a reasonable method of implementation. These devices usually consist of levers, plungers, and springs and are normally *user implemented* by being built into the system to be automated. Typical mechanical logic elements are illustrated in Figure 5.1.

The logic of the NOT device is obvious once the forward position is defined as ON, and backward as OFF. If the input is ON, then by the nature of the lever, the output is OFF. The other devices are almost as simple, but some explanation is warranted.

Two conditions must exist in the case of the AND device. The operating force of the spring must exceed the force associated with either the displacement at A or B, but must be less than their total combined force; and the lever on which the two displacements operate must be freely pivoted. Similarly, the spring force on the OR device must be less than the lesser of the forces from the displacements at A or B, and the crossbar must be rigidly attached to the plunger.

The *MEMORY* device uses the principle of the *over-center spring* to provide a retaining force or detent action. If the spring is more compressed in the center than it is at either end of its travel, a force will be applied that will tend to resist motion away from either extreme position. It will, there-

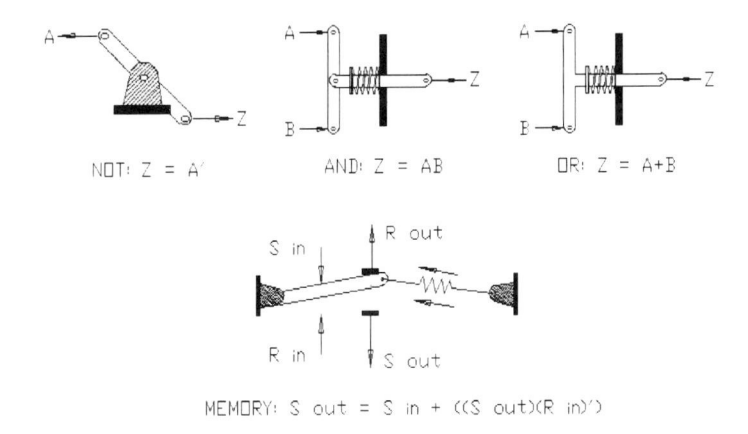

NOT: $Z = A'$ AND: $Z = AB$ OR: $Z = A+B$

MEMORY: $S\ out = S\ in + ((S\ out)(R\ in)')$

Figure 5.1 Mechanical Logic Implementation

fore, *remember* the last input, either S or R. This device is commonly seen in the reversing mechanism of a surface grinder table.

5.3 MECHANICAL INPUT, ELECTRICAL OUTPUT SYSTEMS

This category of devices consists of various mechanically operated electrical switches. Basic switches can be grouped by the number of input positions (single or double throw) depending on whether there are one or two operated positions away from a neutral position, the number of circuits switched (single or multiple pole), whether the electrical contacts in the unoperated position are open or closed (normally open or normally closed), and the nature of the input (momentary or maintained). For example, a single-pole single-throw normally-closed (SPSTNC) momentary switch provides a simple NOT operator, since electric current will flow only if the mechanical actuator is held in the normal or OFF position.

If a number of single-pole single-throw normally-open (SPSTNO) momentary switches are interconnected in a series circuit, they will perform an AND function, since current will flow only if all the switches are actuated. If the same switches are interconnected in a parallel circuit, the OR function will be realized since any one or a combination of input actuations will result in a current flow.

If a single-pole double-throw (SPDT) maintained switch is used, the MEMORY function will be implemented, since the input remains as it was last operated, and the electrical output follows the input. Switching logic is illustrated in Figure 5.2.

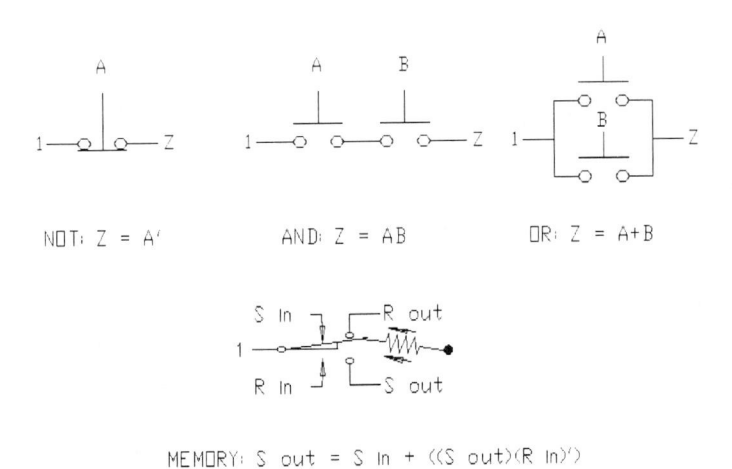

NOT: $Z = A'$

AND: $Z = AB$

OR: $Z = A+B$

MEMORY: $S\ out = S\ In + ((S\ out)(R\ In)')$

Figure 5.2 Switching Logic Implementation

5.4 ELECTRICAL INPUT, ELECTRICAL OUTPUT SYSTEMS

These devices are commonly called *relays*. Relays can vary in size and design from extremely small, sensitive, and fast solid-state devices; through low to moderate power reed relays; to large heavy duty power relays. Some relays that are commonly used in automation applications are illustrated in Figure 5.3.

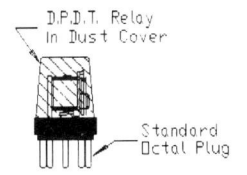

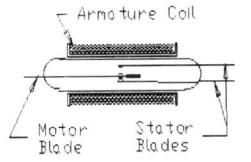

Figure 5.3a General Purpose Relay

Figure 5.3b Reed Relay

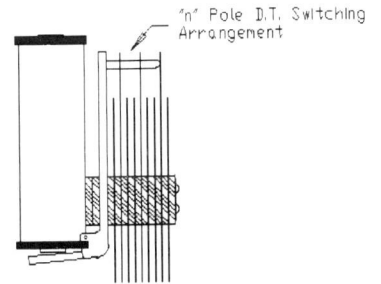

Figure 5.3c Long Frame "Telephone" Type Relay

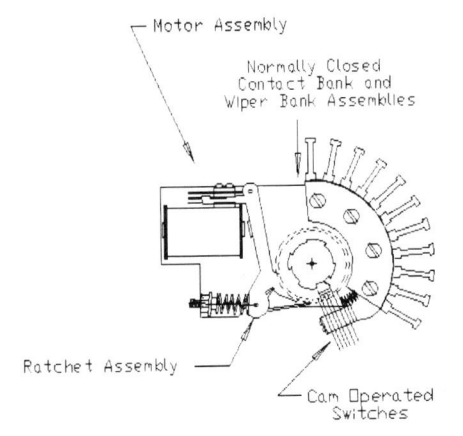

Figure 5.3d Rotary Stepping Switch

The most commonly used electro-mechanical device in automation applications is probably the General Purpose relay shown in Figure 5.3a. It is readily available from many sources, is relatively immune to normal industrial contaminants due to its dust cover, is easily replaced in the case of failure, and is comparatively inexpensive. It is also the most subject to in-service failure. For this reason the plug-in feature is desirable. In spite of this limitation, this type of relay has met with great success in many applications.

The second relay to be considered is the Long Frame Telephone Type illustrated in Figure 5.3c. This particular device has a long history of reliable service in some of the older telephone switching systems. It is extremely flexible in application; as many as twelve single-pole double-throw switching elements can be contained in one device. As stated before, it is quite reliable when properly separated from the environment. Its major disadvantages are the expense of installation and replacement, the limited number of vendors from whom it can be purchased, and the cost; however, when reliability at moderate switching speeds is the dominant factor, it should be considered as a viable device.

Another Telephone Type relay is the Rotary Stepping Switch shown in Figure 5.3d. This relay is the one that was used for many years in the highly successful and extremely reliable *Strowger* system built by Automatic Electric Co., and distributed by General Telephone Co.. This particular device incorporates memory and switching capabilities in one unit. In effect, it enables the simple construction of passive memory sequential systems, which will be described in a later chapter. These devices share the extremely high reliability of the other Telephone Type relays, but suffer from the problems stated above.

A relatively recent addition to the electro-mechanical control and automation devices is the *reed* relay, shown in Figure 5.3b. These relays consist of two parts: the reed itself containing the switching blades in a controlled environment within an encapsulated vial, and an armature coil consisting of one or two windings. The reed is available in either single-pole single-throw normally open or closed configuration, single-pole double-throw devices that are "biased" toward one or the other of the stator blades, or single pole double throw *over-center* devices that can act as memory elements. In a reed relay, when the armature coil is energized, a magnetic field is created. This in turn acts on the motor blade causing a change in position, which in turn enables the switching. These devices vary in size from under one inch to 2 inches in length. Their switching rates approach that of electronic devices, due to the relatively high forces acting on, and the low mass of the moving elements. Single wound coil devices can be used to construct all the logic elements shown in Figure 5.4, double wound coils can be used to construct AND elements as well as other logic systems.

A more extended discussion of the nature and specifics of these types is beyond the scope of this text and is unnecessary since they all operate in essentially the same manner. If an electrical current is present at the input to the device, an output circuit will change state, (open to closed, closed to open, or transferred).

Sec. 5.4   Electrical Input, Electrical Output Systems   61

Relays can be categorized in a manner very similar to that used for electrical switches, which is not surprising since the basic difference between the two groups is the nature of the input. There are, however, two features that are inherent in relays that should be discussed.

The first is the fact that specific provisions must be made for a mechanical or electro-mechanical interlock, or latch, for maintained operation. By the nature of their design, all relays are momentary actuation; that is the output will remain in a changed state only as long as the input is ON. Even the MEMORY type reed relay requires a special over-center configuration to work!

The second feature is the availability of multiple electrical inputs in a single relay, enabling many of the logic operations to be contained within a single device through the use of judicious design procedures, rather than depending on the interconnection of the devices in series or parallel circuits.

The implementation of the logic operators is illustrated in Figure 5.4. Note specifically the manner of connecting the inputs to the relays in order to obtain the desired action, the use of a supply voltage (shown as a 1 signal) to provide power for a NOT output, the use of a latching device to realize a MEMORY, and the ability to construct the universal operator INHIBIT.

The illustrations are drawn using standard *attached* symbols. In all cases, either the A or the SET input is present at the coil. The other coil terminal is connected to *ground* (in telephonic applications, "ground" is the

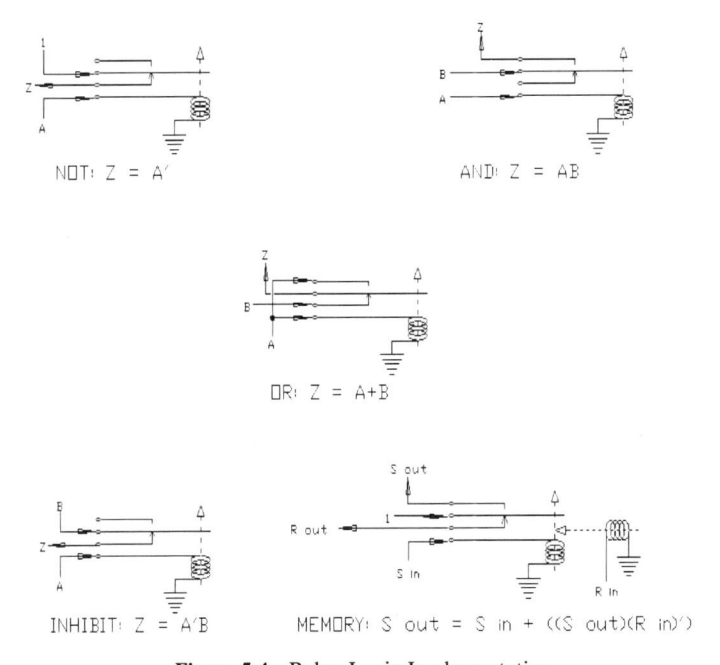

NOT: $Z = A'$

AND: $Z = AB$

OR: $Z = A+B$

INHIBIT: $Z = A'B$

MEMORY: $S\ out = S\ in + ((S\ out)(R\ in)')$

Figure 5.4   Relay Logic Implementation

higher potential positive pole). The switching or blade terminals are shown in the SPDT configuration enabling the transfer of a current flow from one terminal to the other. The movable or motor blade is mechanically or electrically connected to the coil armature; so it moves to break one contact and make another. The timing of this break-make cycle, and its nature (make-before-break or more commonly break-then-make) is critical in many cases and can be specified by the user. The fixed or stator blades make contact with the motor blade to generate the desired switching.

While it was not physically necessary, all the illustrations are of single-pole double-throw (SPDT) transfer devices. The SPDT device was chosen since this particular configuration permits all the logic operators to be constructed from the same piece of hardware and eliminates the need to stock four different configurations in inventory. Economies of scale can be realized by using larger quantities of a single item rather than distributing the same total quantity over four different items, and maintenance is much simplified by minimizing stock variations.

The essential operators and the INHIBIT function are easily traced through the illustrations. This is left to the reader as an exercise in logic. The MEMORY device is also relatively straightforward, but the latching mechanism can cause significant problems. An electro-mechanical interlock, as shown in Figure 5.4, is the most desirable since power is being used only at the moment of SET or RESET, but this effectively doubles the cost of the device since a second operating coil is required. Electrical latching is also possible. The coil current is maintained using a second set of contacts on the relay, and *is* RESET using either a switch or a second relay. The problems with electrical latching are the power consumption and the ensuing coil heating which significantly shortens the service life of the device, as well as the additional cost for the RESET relay.

5.5 FLUID POWER SYSTEMS

Fluid power, or *moving part logic* (MPL) devices are normally available as spool valves, poppet valves, and diaphragm or multiple diaphragm valves. They can be actuated by means of mechanical displacements, electrical currents, or fluid pressures, among others. There are as many different designs as there are manufacturers. They are widely applied to automation because of this flexibility in actuation, as well as their insensitivity to noise and the environment, and their high reliability and maintainability.

A basic valve, and one of the most flexible in adaptability and application, is the three-way spring-returned spool valve. A typical spool valve and its symbolic representation are illustrated in Figure 5.5. A brief discussion of the operation of this and the four-way valve, shown in Figure 5.6,

Sec. 5.5   Fluid Power Systems   63

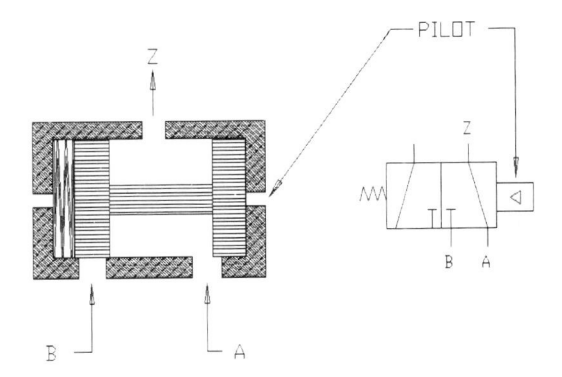

Figure 5.5a Cross Sectional View   **Figure 5.5b** Schematic Diagram

Figure 5.5 Three-Way Spring Return Spool Valve

follows; the applicable logic configurations will be shown in Figure 5.9 at the end of this section.

Consider the valve in Figure 5.5. Under neutral (rest) conditions, the spring acts on the end of the movable member (spool) and forces it to the extreme of its travel as shown. In this orientation, the input A is directly connected to the output Z through a circular groove in the spool itself. The input B is blocked by the spool, thereby preventing any flow from this port to the output. This is called a blocked port.

If an actuating force exceeding the spring force is applied to the right end of the illustrated spool, the spool will travel to the extreme left position. The spring will be compressed but will not develop sufficient force to overcome the actuating force, and the spool will remain in this position as long as the actuating force is maintained. In this orientation, the port B is connected to the output Z, and port A is blocked. When the actuating force is released, the spring once again forces the spool to the right, and the original conditions exist.

The second basic spool valve is the four-way valve shown in Figure 5.6. This valve differs from the one previously discussed in that it has two inputs, A and B; and two outputs, Z_1 and Z_2. It also uses externally applied forces to shift the position of the spool rather than a spring return. The spool moves from either extreme position to the other, and remains in either of the extreme positions due to friction or the presence of a mechanical or electrical detent. If a force is applied at the left end of the spool, connections exist between A and Z_1, and between B and Z_2. If, however, the force is applied to the right end, then port A is connected to Z_2 and B to Z_1.

In addition to spool valves, ball and/or poppet and diaphragm valves are used as control elements in automation systems. These devices are illustrated in Figure 5.8.

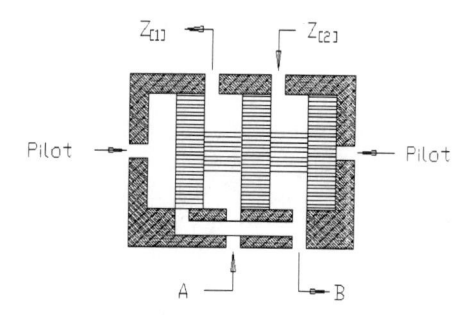

Figure 5.6a Cross Sectional View

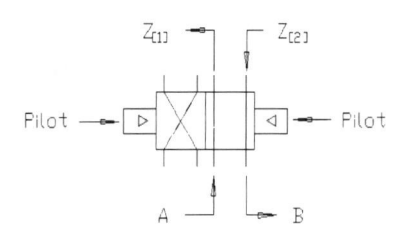

Figure 5.6b Schematic Diagram

Figure 5.6 Four-Way Double Piloted Spool Valve

Before undertaking any further discussion of logical automation devices, it is appropriate that the physical as well as the logical nature of an *OR* device be considered.

In addition to its use as a logic operator, it is essential that an *OR* be used when combining inputs, or *fanning-in*. Consider the two-input, one-output block shown in Figure 5.7a. Further, consider that ports *A* and *B* will always receive input signals from some source, and that port *Z* will always connect to the next device, or load. With this arrangement, there are three possible functional combinations which are shown in Figures 5.7 b–d.

Figure 5.7b shows the case where the unused input is blocked. In this event, the passage connecting to this port might just as well not be there. However, consider what would happen in the event that this port is not blocked, but rather vented to the atmosphere, as is usually the case.

If there is a load at *Z* and a vented (grounded) port at *A*, the maximum pressure (voltage) in the system must be that of the vent (ground). Under these conditions, the output potential is the same as the vented (grounded) port, or 0. By similar reasoning, the situation illustrated in Figure 5.7c represents the same conditions.

Figure 5.7d shows the situation when both the inputs are energized. In this case, the output would be 1. Therefore, for all practical purposes, the fan-in condition does not provide an OR function when used with vented inputs, but rather an AND. In the case of unvented or blocked input ports the component is in reality a fan-out device, since the unenergized input becomes another output.

Sec. 5.5   Fluid Power Systems   65

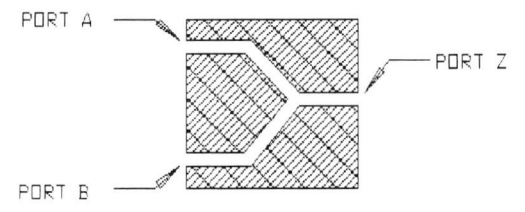

Figure 5.7a 2 Input, 1 Output Block

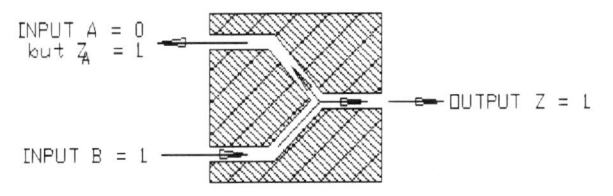

Figure 5.7b B = 1, Z = B, A = B

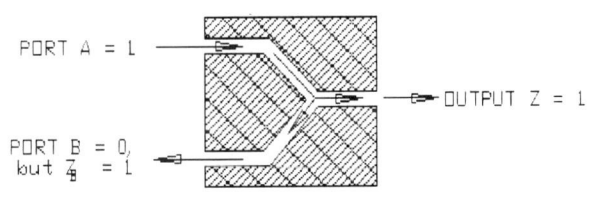

Figure 5.7c A = 1, Z = A, B = A

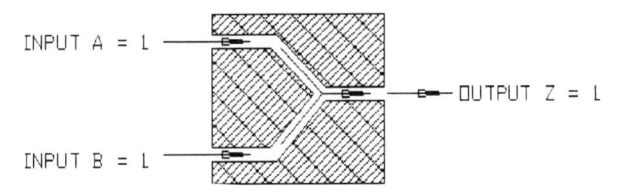

Figure 5.7d A = 1, B = 1, Z = AB = 1

Figure 5.7 The OR/Fan-In Dilemma

Figures 5.8a–c show diagrammatic and schematic views of the primitive ball or poppet valve in an OR configuration. Attention should be paid to the fact that these are true ORs and not fan-in elements. These devices are in reality fluid-power diodes in that they not only provide a true logical *OR*, they also permit flow in only one direction. Other logical configurations can be constructed through a combination of multiple poppets, poppets and spools, and poppets and diaphragms (see Figure 5.8e), and are available from a number of manufacturers.

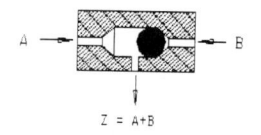

Figure 5.8a Ball Check Valve Cross Section

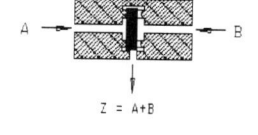

$Z = A+B$

Figure 5.8b Internal Poppet Valve Cross Section

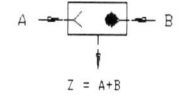

$Z = A+B$

Figure 5.8c Schematic Diagram

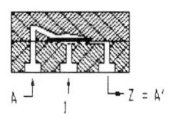

$Z = A+B$

Figure 5.8d Single Diaphragm Valve NOT Configuration

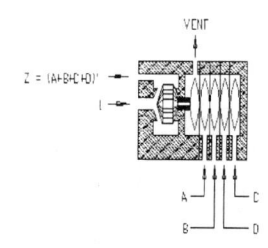

$Z = A'$

Figure 5.8e Multiple Diaphragm Valve NOR Configuration

$Z = (A+B+C+D)'$

Figure 5.8d illustrates a NOT device using a diaphragm valve. In the absence of a pressure at A, flow passes from the energized input 1 to the output Z. If a pressure less than 1 is present at A, the diaphragm moves downward due to the greater area above the diaphragm than below it. These devices are not only logical; they also act as slightly amplifying repeaters, since the output energy level is greater than the input signal associated with that output.

A unique valve available from Festo in West Germany deserves special consideration and is shown in Figure 5.8e. A signal present at any one or a combination of the logical inputs will move the poppet to the left, thereby shutting off the flow of air to the output Z. These devices are not readily available in the United States, but can be obtained through selected distributors.

Fluid power moving part logic devices are available in standard logic configurations and in many specialty arrangements. In addition to the classical NOT, AND, OR, and MEMORY devices, there are systems that offer NOR, INHIBIT, and other proprietary devices, as described above.

One of the more significant limitations of most MPL devices (other than the multiple diaphragm valve) is that, in most cases, there can be no

Sec. 5.5 Fluid Power Systems 67

more than two inputs to any single device. In fact, in the event that there are n inputs required to a given gate, then the greater of one or $n - 1$ devices will be required for that gate.

In keeping with the previous examples of logic implementation, it is sufficient to show that these operators can be realized. The most flexible devices will be used in this demonstration for the same reasons as stated earlier, namely economies of scale.

The most primitive logic element is the NOT, shown in Figure 5.9, implemented through the use of a 3-way spring return pilot (pressure actuated) valve. Since an output signal is required when the input is OFF, an active device is needed. The external source of power is indicated by the use of the 1 in a manner similar to that used in earlier discussions. It can be readily seen that if no signal is applied to input A, an output signal Z is present, but if input A is energized, the output is OFF. The other operators are equally as straightforward.

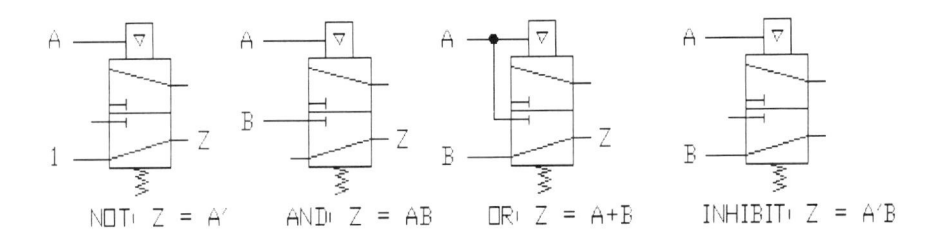

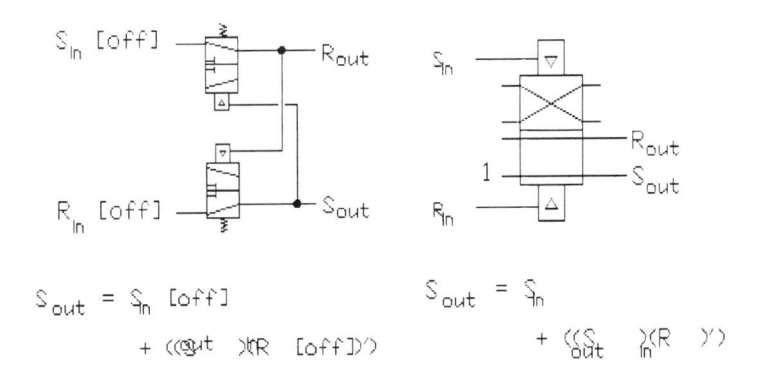

NOT: $Z = A'$ AND: $Z = AB$ OR: $Z = A+B$ INHIBIT: $Z = A'B$

MEMORY Implementation

Figure 5.9 MPL Logic Symbols

Note that in all cases except the last MEMORY implementation, single piloted 3-way valves have been used to implement the logic. In all cases, except the two element MEMORY device, these have been arranged so that when an input signal goes ON, the component functions as described. In the case of the two element MEMORY, two INHIBIT elements are used, with their inputs arranged so that the MEMORY functions when the inputs are turned OFF. By doing this, NOT gates at the inputs are eliminated.

In addition to the two INHIBIT configured MEMORY, a double piloted 4-way valve is used to realize the MEMORY operator. It should be evident that a single valve can be used to generate all logic operators, but in the case of MEMORY this is not an efficient use of resources.

5.6 FLUIDIC AND ELECTRONIC SYSTEMS

Fluidics are to fluid power systems what electronics are to relay systems. They have similar characteristics as well as some individual features.

Fluidic elements can be characterized by the fact that they switch or divert the flow of a fluid (normally, but not always, air) solely by fluid dynamic phenomena. Electronic elements can be characterized by their diverting the flow of their medium (current flow or electricity) solely by electronic phenomena.

The use of both fluidics and electronics result in low power systems. Fluidic systems are sensitive to contamination of their working medium. Electronic systems are sensitive to spurious signals (noise).

With proper conditioning of the working medium, fluidic systems are essentially failure free, since there are no moving parts and consequently no friction. The two factors that must be considered when conditioning the air are particulate contamination and moisture. Particles can be filtered out, but this requires an extremely fine filter which develops a significant pressure loss quite rapidly. This means that constant surveillance is required. Moisture can be removed by several schemes, but this removal is expensive and energy inefficient. Several fluidic elements are shown in Figure 5.10.

Fluidic devices are essentially proportional amplifiers, but through judicious design, their transition from full flow to extremely low flow can be made highly non-linear. These non-linear, sharp-cutoff devices can be used in logical automation applications.

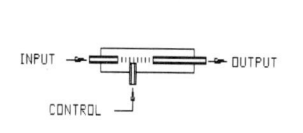

Figure 5.10a Simple Turbulence Amplifier

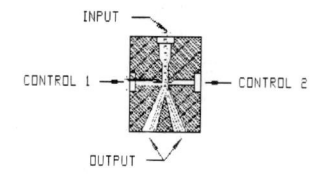

Figure 5.10b Proportional Amplifier

Sec. 5.6   Fluidic and Electronic Systems   69

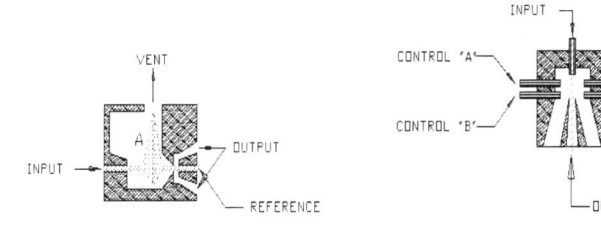

Figure 5.10c Impact Modulator

Figure 5.10d NOR Element

A simple turbulence amplifier is illustrated in Figure 5.10a. If a laminar flow source is present at the input, then the addition of a signal at the control will transform this laminar flow to one that is turbulent, thereby sharply reducing both the pressure and flow present at the output, creating a NOT device. Figure 5.10d shows the application of this phenomenon to the design of a fluidic NOR element.

Figure 5.10b illustrates a device where the relative amount of flow in the two outputs is directly proportional to the relative flows at the controls 1 and 2. Through careful design, the nature of which is not appropriate to discuss here, these flows will *attach* to the outer wall due to what is known as the *Coanda effect*, and remain there until the opposite control is energized. This results in a MEMORY or BISTABLE device.

An impact modulator is shown schematically in Figure 5.10c. This may be the only fluidic device that is currently available as a stocked item. It is used in air conditioning control systems by Johnson Controls and, when properly designed and applied, provide very linear output. It is rarely used in digital systems but deserves mention if for no other reason than its availability.

The nature of the operation of an impact modulator is a function of the relative low level flows at the input and the control ports of the device. These are adjusted to provide *impact* in the region shown, which passes out the port marked VENT, resulting in little if any flow at the annular output. As the flow of the input increases, the region of impact shifts to the right, where some of the flow exits through the output.

Electronic control systems are extremely fast. Switching speeds in the neighborhood of 10^{-6} to 10^{-9} seconds are normal. If electrical noise is present in the environment, signals that would be ignored by systems with slower responses will trigger a response just as if they were a normal input. The usual source of this noise is the alternating current power normally used in the American industry.

There are two common schemes used to solve the noise problem. Either the source of the perturbation is removed by the use of direct current devices and filters, or extensive shielding or electronic isolation is used to protect the electronics from its environment.

In both fluidic and electronic systems, sources of regular repetitive signals are available. These devices are called *multivibrators* or *clocks*. These clocks permit the use of synchronous sequential control systems. These systems are significantly simpler to design than are asynchronous systems, but they do not result in optimal cost-effective automation. Synchronous control has its place in communications and allied fields. In the case of machine and other industrial automation where the operating cycle is dependent on many factors, event timing (resulting from asynchronous techniques) is desired.

Both fluidic and electronic systems can be made in high density packages, where the nature of the individual logic device is not readily identifiable. This can result in the construction of a system where, in the event of failure, the entire module would have to be replaced rather than just a single element, which can result in rapid system repair. Unfortunately, this technique also results in high material costs.

For all these reasons, fluidic or electronic controllers should be used *only* under the conditions where they are economically viable and where the appropriate maintenance facilities are available. These two conditions normally occur only in large industrial organizations where significant quantities of product are being manufactured.

5.7 OTHER DEVICES

It has been shown that all the required logical operators can be synthesized through the use of different hardware components, depending on the nature of the problem and the desires of the designer. Once the implementing hardware has been selected, it often becomes necessary to condition or *wave shape* signals.

Wave shaping is done by delaying the turning ON of a signal, delaying the turning OFF of a signal, and turning a signal OFF after a specific period of time. These are called DELAY-IN, DELAY-OUT, and ONE-SHOT (monostable), respectively.

These wave-shaping elements and their implementation using relays and MPL devices are illustrated in Figure 5.11. (Similar functions can of course be generated using other systems.)

Note that all these wave-shapers are implemented by installing the appropriate resistance-capacitance networks. Consider the relay implemented DELAY-IN device. Upon application of a current at point A, the capacitance is charged at a rate dependent on the setting of the variable resistor. When fully charged, current will flow into the coil winding, thereby actuating the armature blade and closing the output contacts.

In the case of the MPL DELAY-OUT, if a signal exists at A, flow is directed to the pilot which operates the valve. Eventually, depending on the

Sec. 5.8   Programmable Logic Controllers (PLCs)

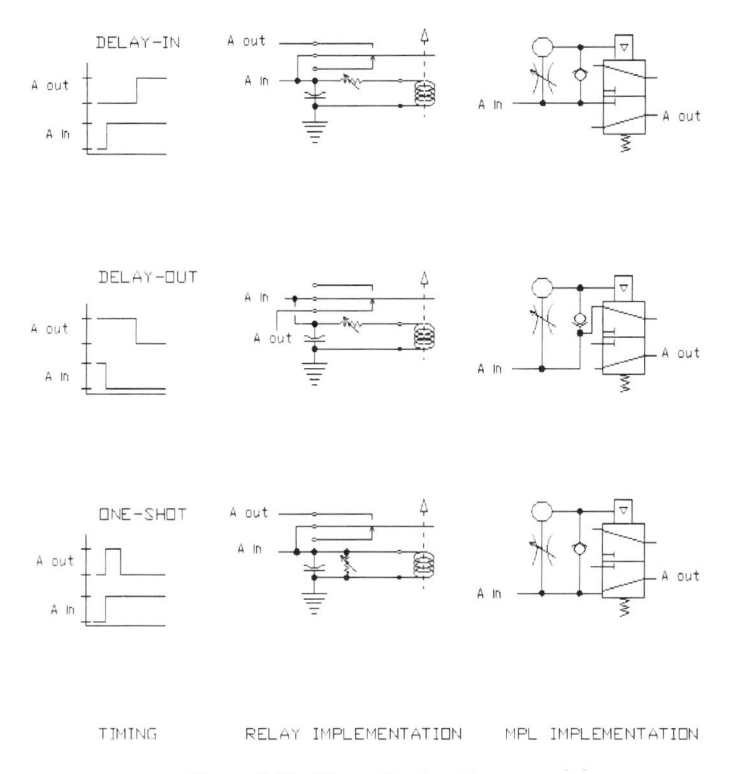

Figure 5.11 Wave Shaping Elements

setting of the variable flow-resistance, the fluid capacitor (chamber) is filled. When the signal is removed at A, exhaust flow is blocked by the check-valve and is dissipated through the chamber. When the pressure reaches a low enough level, the spring returns the spool thereby cutting off the output flow.

The ONE-SHOT is essentially the same as the DELAY-IN with the single exception of the point of connection of the input signal A to the valve body.

5.8 PROGRAMMABLE LOGIC CONTROLLERS (PLCs)

5.8.1 Background

For a number of years, automobile manufacturers automated their major production lines using sensitive switches as input devices to relays, then hard wiring these relays to motor starters and other operators. This was a typical, and reasonable, design approach when large quantities of product were to be manufactured over an extended period of time. Unfortunately, an annual design change usually necessitated a change in the production line, and required either a massive rewiring of existing relays or the scrapping of the existing system and its replacement with another relay system that was designed to control the new line.

There were a number of reasons for this approach. By far the most important was that the renovation of the existing control system cost more than its scrapping and replacement. This was due to the extremely complex nature of the control system itself, coupled with the lack of good documentation of what had already been developed.

In the late 1960s, General Motors, among others, began experimenting with alternate control equipment. Ultimately, a system was developed using solid state devices in a form not unlike the microprocessor that was concurrently being developed for the microcomputer industry.

The symbology that was in common use at the time was *detached-logic* diagramming that had been, and still is, used in the electrical industry for specifying the installation of relays. This type of diagramming, commonly called *ladder diagramming*, is easily interpreted by electricians when installing these electromechanical devices, but is totally inadequate when one tries to logically describe or document the operation of the system. Historical precedence has persisted however, and many of the present day *Programmable Logic Controllers* or PLCs continue this practice. Examples of this technique, and a discussion of its appropriate use, will be presented later.

The PLC itself consists of a *Central Processing Unit* (CPU) and a Memory. The CPU directs the operation of the processor, while the Memory stores information which the CPU controls. Inputs to the PLC consist of relatively low level electrical ON/OFF signals from sensors of all kinds, such as contact, pressure, force, velocity, and magnetic. These input signals are interpreted in the CPU, and the results amplified and directed to appropriate output devices such as solenoids, motor starters, or valves.

The operation of the PLC is *software* based. This means that its control logic function can be changed by reprogamming the Memory. Keyboard programming can be done *on line* which means that the logic functions desired can be readily changed. The only time that hardware modifications are needed is in the event that different or additional input or output requirements are needed.

5.8.2 Comparisons to Other Hardware Systems

PLCs compete with all the other hardware systems: Electro-mechanical relays, MPL devices, solid state relays, and microcomputer based methods.

Relays and MPL devices are high-level devices in that they *can* use input signals of higher power, thereby mitigating the effect of electrical shop noise. They are familiar to both installation and maintenance personnel.

Sec. 5.8   Programmable Logic Controllers (PLCs)   73

Aside from the problems with reprogramming, relays have two major faults compared with PLCs: they are significantly less reliable and their installation is considerably more expensive. However, for smaller systems where changes are not anticipated; relays, especially socket-mounted reed relays or MPL systems, could well be the techniques of choice.

Solid state components are nothing more than electronic versions of the common electro-mechanical relay. In the event that a control current is impressed on the device, an output current will flow, or vice versa. These solid state elements are more reliable than the older electro-mechanical components, but their maintainability on the shop floor is much lower.

Since PLCs, microcomputer-based PCs, and solid state relays are low level devices, they are much more sensitive to spurious noise signals and provisions must be made to insulate these from shop-generated electrical noise. This is *not* a trivial problem. For the most part, this sensitivity to spurious noise has been reduced in the PLC to an acceptable level. Another consideration is the ability of the PLC to be reprogrammed while *on line*. This minimizes production downtime and facilitates changeovers.

5.8.3 Programming PLCs

There are three methods in common use for programming PLCs. These are:

1. Relay *Ladder*, or *detached logic* diagramming
2. *Attached logic* diagramming (discussed in the next chapter), and
3. Microprocessor programming language.

5.8.3.1 Relay *ladder* diagramming. The following brief discussion is intended to familiarize the reader with the basic technique of ladder diagramming. It is not suggested that this is a viable design tool, since it is not! It is useful in expressing a final design to the builder of the system or as the programming medium. It is probably the most widely used method of programming PLCs.

A ladder diagram is nothing more than an electrical wiring diagram showing the interconnection between the input elements, the control components, and the output devices. It is started by constructing two vertical lines which represent the electrical leads with the circuit voltage applied between them. As components are added, horizontal lines are drawn between the verticals.

The symbols used are defined by appropriate standards. However, since each manufacturer of PLCs has its own methodology, further details are inappropriate. Some of the more common symbols are shown below.

In order to illustrate the technique, consider the following example.

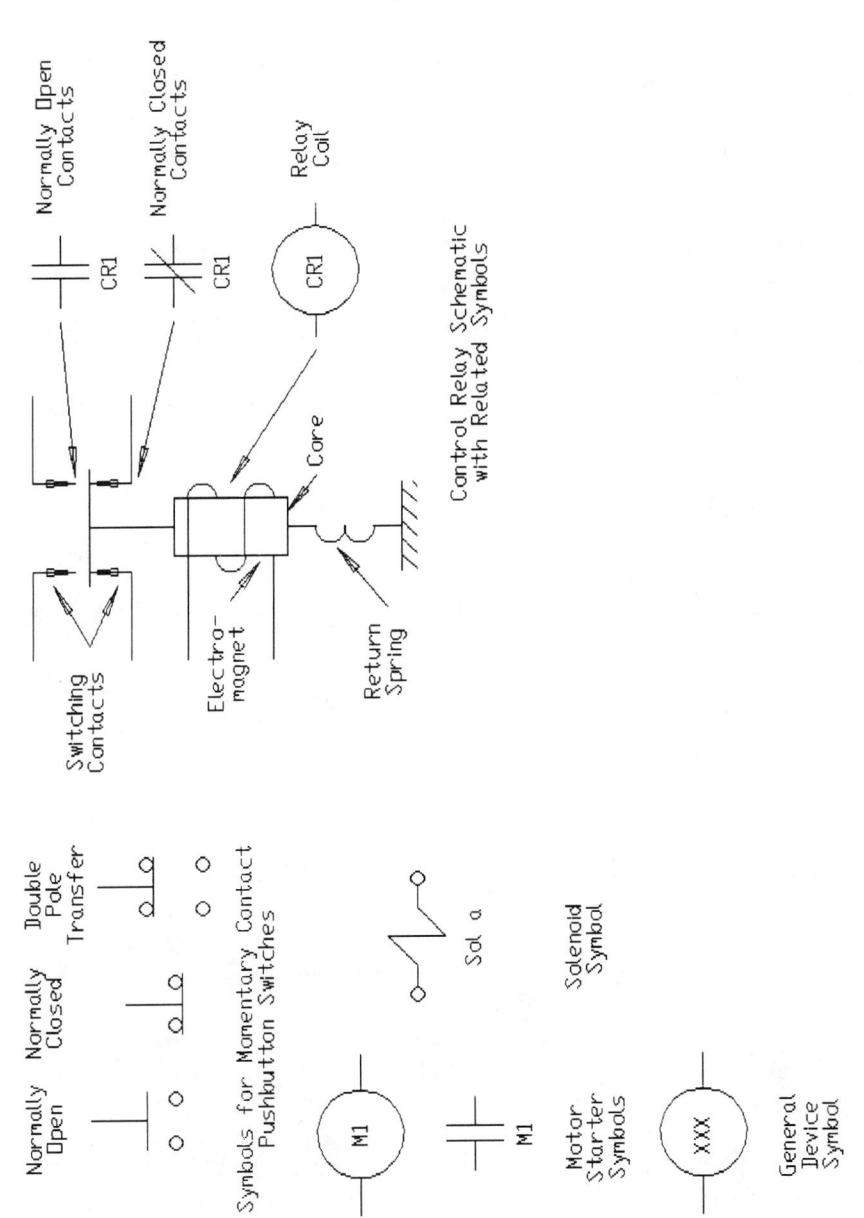

Figure 5.12 Some Electrical Control Symbols

Sec. 5.8 Programmable Logic Controllers (PLCs)

A system employs four motors. The operation of these motors must be as follows:

1. Motor 1 (M1) and motor 2 (M2) cannot both run at the same time.
2. Motor 3 (M3) must run when either M1 or M2 are operating, but cannot operate by itself.
3. Motor 4 (M4) starts only when M1 or M2 (whichever is running) is stopped. M4 will then run for a predetermined length of time and then stop.

M1 is started by operating M1 START. M1 is then latched. Normally closed contacts for M2 are in series with the starter switch in order to prevent M1 and M2 from running simultaneously.

M2 is started by operating M2 START. M2 is latched. Normally closed contacts

$$M1 = Motor\ 1 \qquad M3 = Motor\ 3$$
$$M2 = Motor\ 2 \qquad M4 = Motor\ 4$$

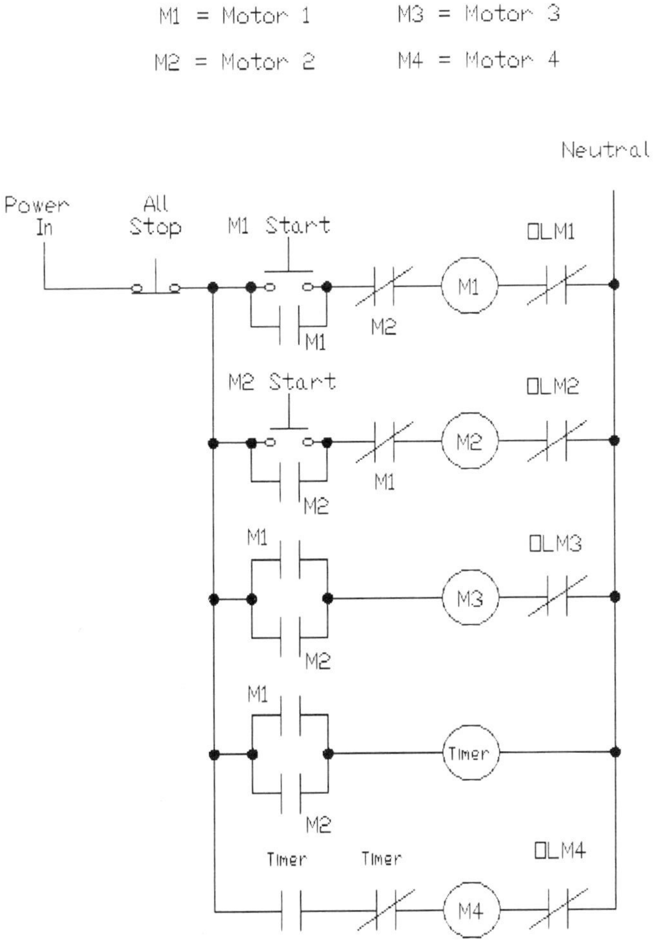

Figure 5.13 Ladder Diagram

for M1 are in series with M2 START to prevent M1 and M2 from simultaneous operation.

Normally open contacts for both M1 and M2 are in series with both a timer and M3. In the event either M1 or M2 is operating, M3 will also run. The timer should also act as a failure detection device; so it will start when it has *no* current imposed on it. Then after a predetermined period of time it will open.

When the timer energizes the first relay associated with M4, it will close, thereby starting this motor. After a predetermined period of time, the normally closed contacts on a second relay will open, thereby deenergizing M4.

5.8.3.2 Attached logic diagramming.

A much more complete discussion on this *design* technique is given in the next chapter. However, it is appropriate that the logic diagram for the above mentioned example be presented for comparison purposes.

The symbology for this *design* method consists, not surprisingly, of three logical devices: AND, OR, and NOT, a MEMORY, and a one-shot device that is used as a timer, as discussed earlier. The major feature of the attached logic diagram is that it is an historical document in that it shows exactly what the system is supposed to do.

Consider the system illustrated in Figure 5.14. Inputs are always on the left while outputs are to the right. Junctions are shown on this diagram

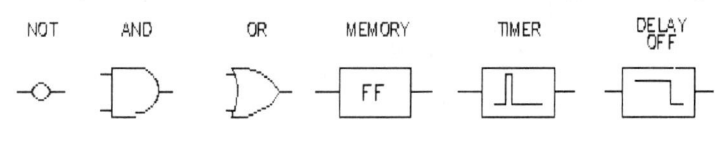

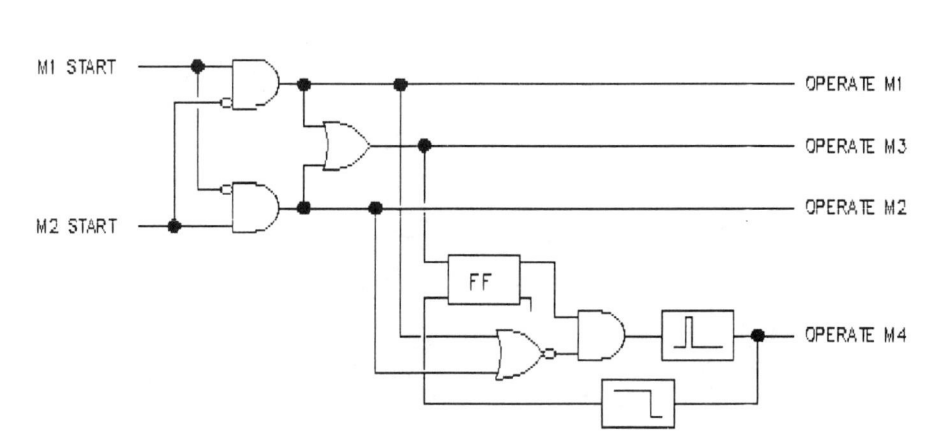

Figure 5.14 Attached Logic Diagram

Sec. 5.9   Selection of Implementing Hardware: Other Criteria   77

by solid dots. These symbols are interconnected as shown above to form the logic diagram.

Consider the output OPERATE M1. The diagram can be read as: operate M1 iff M1 START is operated AND M2 START is NOT operated. Similarly, one could read: operate M2 iff M2 START is operated AND M1 START is NOT operated. Continuing with this process, one can see that M3 will operate iff either M1 is operating OR M2 is operating (or if both M1 AND M2 are operating, but this is precluded by the other logic). Finally, operate M4 iff M1 OR M2 have been running (to set the MEMORY), then NOT (M1 is operating OR M2 is operating). This is an accurate translation of the problem statement.

It should be noted again that this technique permits the documentation of the operation of the system in a much clearer manner than does the ladder diagram. The reader should compare the two to verify this statement. The similarity of relay logic, MPL, and logic diagrams should also be noticed.

5.8.3.3 Programming languages. The third commonly used programming method for a PLC is a *programming* language. Each manufacturer has its own system. Some use a low level language such as a variant of *Assembler*. Other utilize high level languages such as *FORTRAN*. Still others use their own proprietary language.

Since there is such a diversity of languages, further study is recommended as required by whatever system is being implemented.

5.9 SELECTION OF IMPLEMENTING HARDWARE: OTHER CRITERIA

As is stated at the beginning of this chapter, *A logic system has no internal preference regarding its implementation.*

The most significant reasons to select a particular implementation technique are the nature of the system being automated and the environmental conditions under which this system operates.

When one considers the costs involved in the design of a system, there are no significant differences between designing a MPL system or an all electronic one. Implementation costs can vary widely, but component costs are usually a small portion of the total cost and are, therefore, only of moderate interest, *except* in the case of a system that itself will be produced in high quantities or one where reprogramming is a critical consideration. Naturally, in these cases, component, fabrication, assembly, testing, maintenance, and other costs of this nature become highly significant.

Other criteria in the decision as to which hardware plan should be implemented are the efficiency of the devices, their response times, reliability, and their frequency of operation.

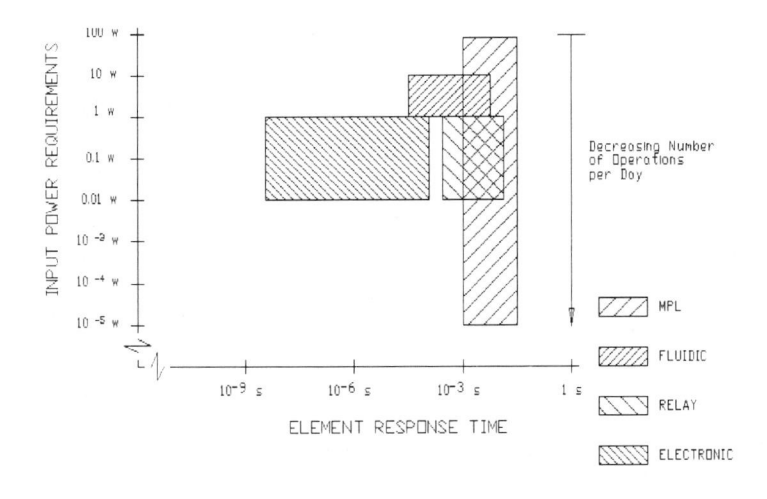

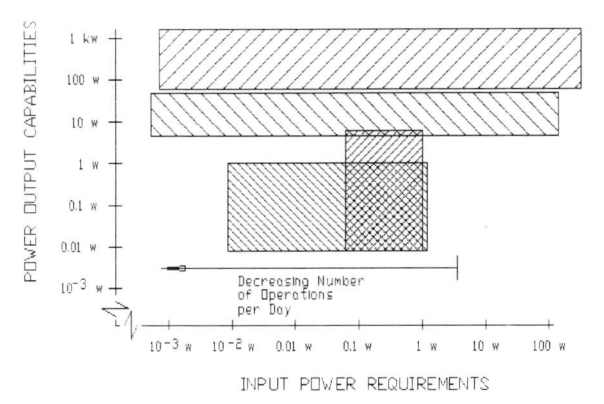

Figure 5.15 Logic Device Comparisons

Device efficiency can best be stated as the ratio of the output power that a device can handle compared to the input power that is required to operate the device. The lower illustration in Figure 5.15 shows this efficiency as a function of the number of operations per day per device. It is readily apparent from this chart that high output/input power ratios can be obtained through the use of either MPL or relay logic systems in the event that the components are not operated very often.

The upper chart shows the relationship of the required input power as a function of frequency of operation compared to element response time. The fastest switching devices are those with the lowest power requirements.

From this data, it is reasonable to draw the following conclusions.

1. If extremely high speed switching or element response is mandatory or reprogrammability is a significant factor, electronic systems are far

superior to any other. This is the case in very high speed sorting systems, data analyzing systems where many comparisons must be made in a short period of time, or large scale automation programs that undergo regular changes as in the automotive industry. In these cases, a programmable microprocessor might be favorably considered.

2. In machine automation, when relatively few logical operations are required, the rate of operation is moderate, and significant power must be handled by the control system, MPL and Relay systems are the most desirable.

3. In the event that extremely hazardous environmental conditions are present, a fluidic automation system might be the most reasonable selection.

4. For all other conditions—the vast majority of automation applications—costs, as discussed briefly in Chapter 1, become the dominant criterion, followed by the individual expertise and likes of the designer.

5. In conclusion, contrary to much popular opinion, the choice of the method of implementation is relatively insignificant when *designing* a logical automation system. To repeat the opening statements of this chapter . . . and add a third

   - An optimal system must do exactly what is required, no more, no less; as well and as reliably as is necessary; at the lowest possible cost.
   - A logic system has no internal preference regarding its implementation.
   - The most successful automation is *always* performed at the lowest possible technological level. Seldom if ever is implementation at the highest (or latest) technological level effective.

5.10 PROBLEMS

1. Design and draw an optimal circuit using mechanical devices *only* for the following control equations:
   a. $Z_1 = AB' + CD + A'C'D$
   b. $Z_2 = AB'C$
   c. $Z_3 = A(B + C) + A'B$
2. Repeat problem 1 using switch logic *only*.
3. Repeat problem 1 using relay logic *only*.
4. Repeat problem 1 using MPL logic *only*.
5. Design and draw an optimal circuit using relay logic *only* for the following control equations. Note that only one *combined* system is required.
   a. $Z_1 = A'B' + B'C' + ABC$
   b. $Z_2 = (A' + B)(C' + D')(A + C + D')$
   c. $Z_3 = E'F + EF' + E'F'$

6. Repeat problem 5 using MPL logic *only*.
7. Repeat Problem 5 using a PLC. Express your design in a relay ladder diagram.
8. Design and draw *optimal* circuits using relay logic *only* for the following systems:
   a. $Z = a'b' + a'c' + c'd' + b'cd$
   b. $Z = a'c' + b'c + acd' + bc'd$
   c. $Z = (c + e + f')(d' + e' + f')(a' + d + f)(b' + c' + d + e + f')$
9. Repeat Problem 8 using MPL devices *only*.
10. Repeat Problem 8 using a PLC. Express your design in a relay ladder diagram.
11. Design and draw an *optimal* control circuit that will perform the following function. Use MPL devices *only*.

    An air conditioning system should be ON in the event the temperature within the building is greater than 75° F, the time is between 7:00 A.M. and 6:00 P.M., and it is a normal workday. It should be OFF under all other conditions unless it is manually turned ON, under which condition the system should operate on the normal workday schedule.
12. Repeat Problem 11 using electro-mechanical devices *only*.
13. Repeat Problem 11 using a PLC. Express your design in a relay ladder diagram.
14. Design and draw an *optimal* control circuit that will perform the following function. Use electro-mechanical devices *only*.

    A telegraphic tape reader used as an input to a communication system has five tracks, each containing a series of 0 and 1 signals represented by the absence or presence of a hole in the tape, respectively. As the tape passes over a series of sensors, the pattern is translated into a logic signal on lines a_1, a_2, a_3, a_4, and a_5. Of the 32 possible signal combinations that can be realized, only two are invalid: 00000 since there is no way to know if this is a space between patterns or a desired signal, and 11111 which is used as an *erase* character.
    a. Generate a logical 1 output if a valid pattern is read.
    b. Generate a logical 1 output if a hole pattern 11011 is read before 01110. Note that this system will require a MEMORY to remember the required prior event.
15. Repeat Problem 14 using a PLC. Express your design in a relay ladder diagram.

Chapter 6

COMBINATIONAL LOGIC SYSTEMS

Combinational logic systems are systems in which the output is not a function of the sequence of activation of the system inputs. In fact, these systems operate as if all their inputs were activated simultaneously. In reality, the simultaneous occurrence of all inputs is highly unlikely. The effect of this will be discussed later in this chapter. Now it is enough to say that when designing combinational logic automation systems, the input signals are considered as if they were simultaneous, and corrections for this *error* are made later in the design process.

These systems require no MEMORY devices since the outputs are determined from the present values of inputs. Because of the nature of combinational systems, the sequence of input events is unimportant to the ultimate output; only the existence of the inputs are significant, within the limitation discussed above. This is not meant to imply that timing of an input is not critical. Because of the nature of various hardware devices, it is sometimes desirable that one signal turns ON before another, or turns OFF after another, or has its ON/OFF cycles shaped with respect to time in some manner. Again, this will be discussed more fully later in this chapter.

Combinational systems comprise a very large segment of the field of automation systems. Sorting systems, inspection systems, and parts feeding systems are all examples of combinational logic automation.

When designing an automation or control system, the dominant design requirement is that *the logic must enable the correct operation of the machine being automated*. Only after the logic design is specified, can the selection of the most desirable hardware system can be made. After the specific devices are selected, timing and signal shaping are performed.

6.1 LOGICAL DESIGN SYMBOLOGY

It has been amply demonstrated that all the essential logic operators can be implemented in many different ways. Since our goal is to design an optimal system, the first step necessary is to state the operating requirements and synthesize these requirements through a set of logical equations. These equations will contain an algebraic representation of the variables and the operators that act on those variables. Once the equations are written, they can be minimized as discussed in Chapter 4.

Given a set of minimized control equations, they are next represented in graphical form using what is known as *attached* diagraming. Separate symbols are assigned to each of the operators. Connections are made from the inputs, through the appropriate operators, to the ultimate output. The reason for using the attached form rather than the more common detached form is that the former accurately models the system being designed, whereas the latter is a convention that has been developed over the years for use with relay installations.

The operations that are required to implement the logic diagram are NOT, AND, OR, INHIBIT, NAND, NOR, and MEMORY (flip-flop or bistable). In addition, wave-shaping devices may be required depending on the choice of hardware. These consist of DELAY-IN, DELAY-OUT, and ONE-SHOT (monostable) elements. These logic symbols are shown in Figure 6.1.

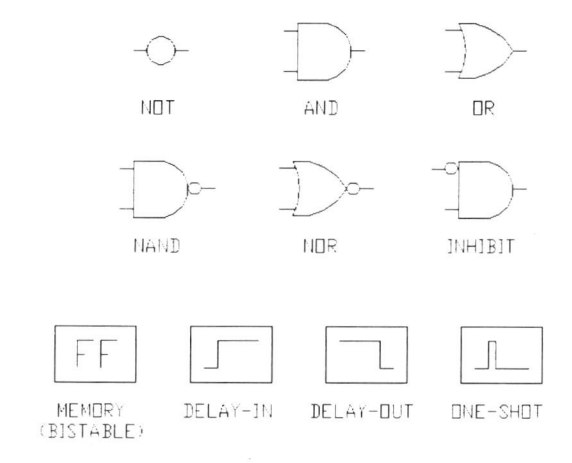

Figure 6.1 Standard Logic Symbols per ASA Y 32.14—1962.

6.2 TYPICAL SYSTEMS

The basic design technique that will be followed throughout all the examples follows.

1. Define the problem so that it accurately represents the desired control system and so it can be expressed in algebraic terms.
2. Optimize the algebraic expression(s).
3. Consider the physical, economic, and fiscal constraints and implement the equations in hardware in the most cost effective manner.
4. Once the implementation has been accomplished, select and design any required wave-shaping.

Consider the system presented in Section 4.1, Example 3:

$$Z = BD + AD' + AB'CD$$

Since there are only four input variables, the mapping method will probably be the most effective technique.

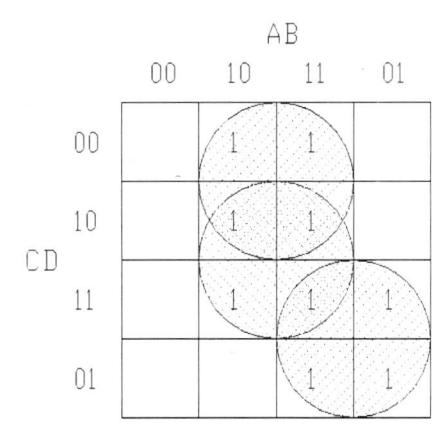

Figure 6.2 Logic Design Map $Z = AC + AD' + BD$.

Note that this is the same form as was previously developed. Note also that there are no more than two inputs to any AND gate and all the required operators are available in either MPL or relay logic. Since this is the case, the logic will be implemented using "AONI" or "AND, OR, NOT, and INHIBIT" techniques, that is, relay or MPL.

Upon study of the final control equation, it is evident that this expression could be implemented by using two ANDs (B and D) and (A and C), one INHIBIT (A and not D), and two ORs operating on the other three terms. This logic has been diagramed and is shown in Figure 6.3.

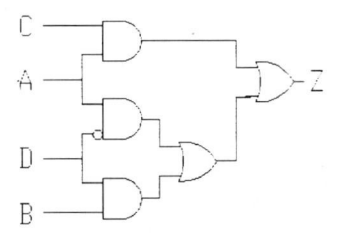

Figure 6.3 Logic Design Circuit (from Figure 6.2).

At this point in the design process, it is essential that we consider the features of the various logic implementation schemes beyond those mentioned in the previous chapter.

The selection decision should be based on the following criteria:

- The nature of the input signal; is it electrical mechanical, pneumatic, or something else?
- The magnitude of the input signal; will it require amplification or is it usable as is?

Sec. 6.2   Typical Systems

- The nature of the environment in which the system will operate; will the presence of an electrical signal pose a safety hazard?
- Is the response time of the control system sufficiently fast compared to the system being controlled so as not to adversely affect it?
- The reliability of the components and the ease with which maintenance can be performed when it is required, remembering the adage " . . . there is never a question as to whether a system will fail, the only question is WHEN! . . . "
- The nature and magnitude of the signals required by the power system; that is, the system that will do the actual work.
- The total cost for the design life of the system; that is the design, component acquisition, construction, operating, and projected maintenance costs.

Some of these factors are outlined in Figure 6.4.

	Electronic	Fluidic	MPL	Relay
Supply Requirement	Precision regulation and filtration		Industrial compressed air	AC or DC electricity
Life Expectancy	Theoretically unlimited		10^7 to 10^9 cycles	5×10^6 to 10^7 contact cycles
Maintenance Requirement	Infrequent; specialist required		Periodic; only basic knowledge required	
Output Level	Very low		High	
Environmental Constraints	Explosion hazards	None	None	Explosion hazards
Response Time (secs.)	10^{-3} to 10^{-8}	10^{-2} to 10^{-3}	10^{-1} to 10^{-2}	10^{-1} to 10^{-2}

Figure 6.4 Selection Criteria for Logic Implementation.

Assume that these factors have been considered. The implementation of the logic circuit in Figure 6.3 is presented in Figure 6.5, where it has been implemented both in MPL and RELAY hardware. Note that these circuits are schematic diagrams of the required hardware and its appropriate interconnections and *are sufficient to specify either system for manufacturing*.

In addition, particular notice should be paid to the one-to-one correlation of the logic diagram shown in Figure 6.3 and the implementations shown in Figure 6.5. This confirms the statement that attached diagraming gives a true representation of a logical automation system!

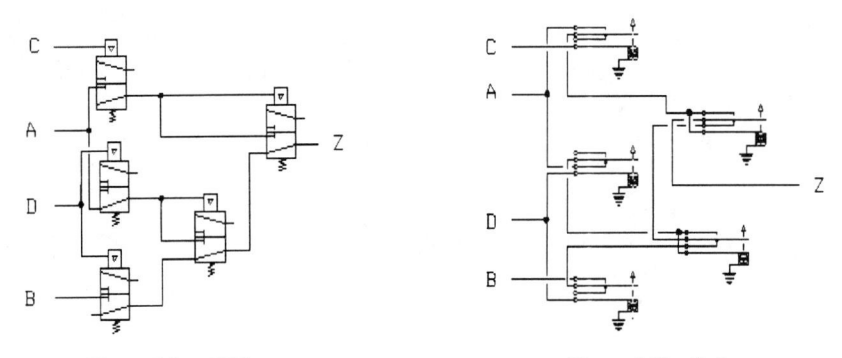

Figure 6.5a MPL.

Figure 6.5b Relay.

Figure 6.5 Logic Implementation Circuit (from Figure 6.3).

Another system that might prove interesting can be described by the equation

$$Z = A'B' + A'DE + B'C + B'CDE + AB'C$$

The optimization for this example will be done using the prime implicant technique. By inspection of this expression, it should be noted that the last three terms reduce to $B'C$, that is, $B'C(1 + DE + A)$, and the entire expression becomes simply $Z = A'B' + B'C + A'DE$. This can be easily implemented using AONI techniques. The truth tables and prime implicant tables are shown in Figure 6.6.

```
      a b c d e
     ---------------
 0 |  0 0 0 0 0  '
 1 |  0 0 0 0 1  '
 2 |  0 0 0 1 0  '
 4 |  0 0 1 0 0  '
 3 |  0 0 0 1 1  '
 5 |  0 0 1 0 1  '
 6 |  0 0 1 1 0  '
20 |  1 0 1 0 0  '
 7 |  0 0 1 1 1  '
11 |  0 1 0 1 1  '
21 |  1 0 1 0 1  '
22 |  1 0 1 1 0  '
15 |  0 1 1 1 1  '
23 |  1 0 1 1 1  '
     ---------------
```

Figure 6.6a Truth Table.

Sec. 6.2   Typical Systems

```
        a b c d e
--------------------
 0  1 : 0 0 0 0 - '
 0  2 : 0 0 0 - 0 '
 0  4 : 0 0 - 0 0 '
 1  3 : 0 0 0 - 1 '
 1  5 : 0 0 - 0 1 '
 2  3 : 0 0 0 1 - '
 2  6 : 0 0 - 1 0 '
 4  5 : 0 0 1 0 - '
 4  6 : 0 0 1 - 0 '
 4 20 : - 0 1 0 0 '
 3  7 : 0 0 - 1 1 '
 3 11 : 0 - 0 1 1 '
 5  7 : 0 0 1 - 1 '
 5 21 : - 0 1 0 1 '
 6  7 : 0 0 1 1 - '
 6 22 : - 0 1 1 0 '
20 21 : 1 0 1 0 - '
20 22 : 1 0 1 - 0 '
 7 15 : 0 - 1 1 1 '
 7 23 : - 0 1 1 1 '
11 15 : 0 1 - 1 1 '
21 23 : 1 0 1 - 1 '
22 23 : 1 1 0 1 - '
```

Figure 6.6b   Merging Diagram.

```
          a b c d e
-------------------------
 0   1   2   3 : 0 0 0 - - '
 4   5   6   7 : 0 0 1 - - '
20  21  22  23 : 1 0 1 - - '
 3   7  11  15 : 0 - - 1 1
```

Figure 6.6c   Merging Diagram.

```
                              a b c d e
-----------------------------------------
 0  1  2  3  4  5  6  7 : 0 0 - - -
 4  5  6  7 20 21 22 23 : - 0 1 - -
```

Figure 6.6d   Merging Diagram.

```
       0   1   2   3   4   5   6   7  11  15  20  21  22  23
--------------------------------------------------------------
* A :  [x][x][x](x)(x)(x)(x)(x)
  :
* B :                  (x)(x)(x)(x)              [x][x][x][x]
  :
* C :  (x)                              (x)[x][x]
```

Figure 6.6e   Prime Implicant Table.

Figure 6.6   Determination of Optimal Form for $Z = A'B' + A'DE + B'C + B'CDE + AB'C = A'B' + B'C + A'DE$.

By selecting B' as one of the input literals, a logic circuit can be constructed using two INHIBIT $((A'D)$ and $(A'B'))$, two AND $((B'C)$ and $((A'D)(E)))$, and two OR $(((A'B')$ OR $(B'C))$ and $((A'DE)$ OR $(A'B' + B'C))$ elements; as shown in Figure 6.7a. This is the direct implementation of the logical expression in disjunctive form.

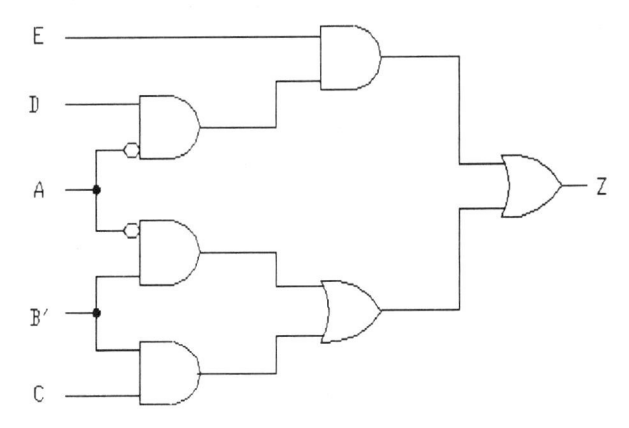

Figure 6.7a   Logic Circuit (from Figure 6.6) $Z = A'B' + B'C + A'DE$.

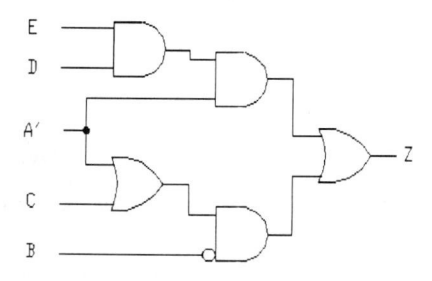

Figure 6.7b Logic Circuit $Z = B'(A' + C) + A'DE$.

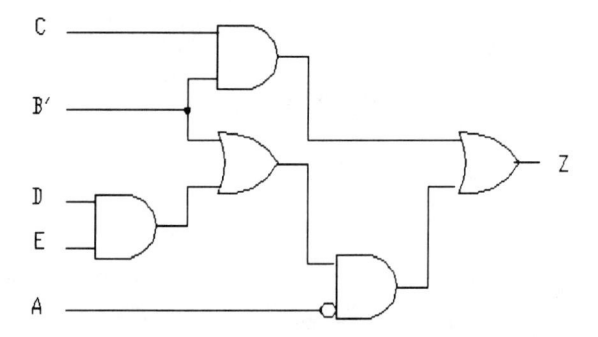

Figure 6.7c Logic Circuit $Z = A'(B' + DE) = B'C$.

Figure 6.7 Logic Circuit Design.

At this point in the process, some of the artistic or nonanalytic elements of design must be considered. These are the intuitive factors that rely so heavily on experience and the ability to instinctively recognize when a form is probably nonoptimal. The results of these intuitive feelings may or may not bear fruit, but one finds that they must be explored. To paraphrase a statement made earlier, the question is never " . . . should these avenues be explored ?" but rather " . . . when is it appropriate that they stop being considered?" The answer to this latter question is very complex and requires significant experience in the design process.

The primary technique, if one exists in this approach, is the algebraic manipulation of the disjunctive form into a *mixed* form that is neither disjunctive nor conjunctive, but is more compact. For example, the statement now being considered:

$$Z = A'B' + B'C + A'DE$$

could be expressed equally as well as either:

$$Z = B'(A' + C) + A'DE$$

or as $Z = A'(B' + DE) + B'C$

Sec. 6.2 Typical Systems

in mixed form. Further study does not produce any *interesting* results, so it is decided to stop with these three statements. The two latter forms are diagramed in Figures 6.7b and c. It can readily be seen that either is more optimal than the original statement since one less operator is required. However before considering the implementation of these circuits, it is appropriate that timing requirements and *hazard* prevention be addressed.

6.2.1 Combinational System Hazards

A hazard is defined as a possible or actual aberration in the output of a logical operator that occurs during the transition from one state to another. These hazards often result in unpredictable operation of a control system. The elimination of hazards presents significant problems for the systems designer. Unfortunately, they result in systems that are somewhat less than optimal from a purely logical standpoint, but for the sake of reliability of operation, they must be done.

Hazards can result from imperfections in automation system hardware even though the system seems to accurately reflect the control equations. This is largely due to the finite switching speed of logic devices, and the fact that input states do not really change simultaneously. It is assumed that a signal is either ON or OFF and there is no transition period between the two allowable states.

The output of a combinational system is solely dependent on the input states. If an input is undefined *during its transition*, the resulting output is undefined as well. The uncertainty is temporary. Unfortunately, it may or may not affect the operation of the machine being controlled.

Hazards in combinational systems are classified as being either static or dynamic; see Figure 6.8. The static hazard can exist when there is to be no change in output, either in the 1 state transition as shown, or in the comparable 0 state transition. These hazards are characterized by an even number of changes in the output state. A dynamic hazard can exist when a

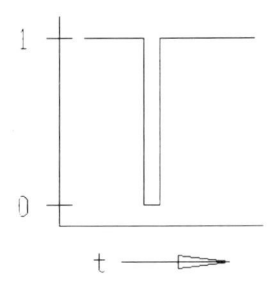

Figure 6.8a Static Hazard. 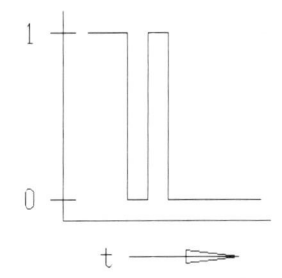 **Figure 6.8b** Dynamic Hazard.

Figure 6.8 Types of Hazards.

change in output state is desired. Odd numbers of output state changes characterize this hazard.

A static hazard can be illustrated by considering the system shown in Figure 6.9. Note that the output of the Karnaugh map is:

$$Z = AB + B'CD$$

Consider the network diagram of Figure 6.9b. Assume that the initial input state is $ABCD = 1111$, and the next state is $ABCD = 1011$. These two input states have exactly the same output state. Theoretically, the output Z should remain ON during the transition, but it is possible that the B signal at the INHIBIT element turns OFF after the B signal at the AND element. If this is the case, then both inputs to the OR element would be momentarily OFF. The output Z would then turn OFF and then ON. This particular hazard results in a change of state 1 and is called a static-1 hazard. If a hazard results in a change in state 0, it is called a static-0 hazard. Static-0 hazards occur when a control equation is implemented conjunctively and the output is in the OFF or 0 state.

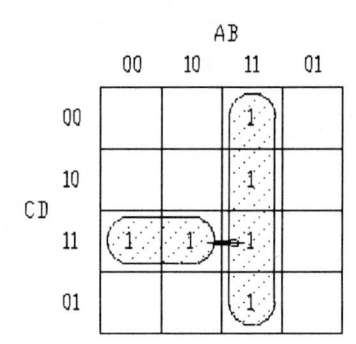

Figure 6.9a Karnaugh Map.

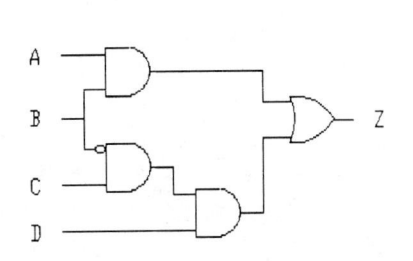

Figure 6.9b Logic Circuit with Hazard.

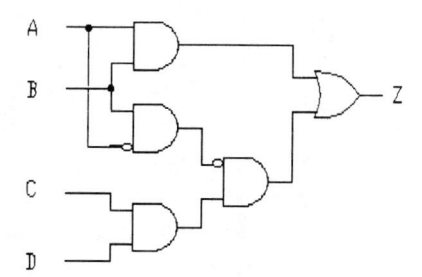

Figure 6.9c Hazard Free Implementation.

Figure 6.9 Static-1 Hazard Elimination.

Sec. 6.2 Typical Systems　　　　　　　　　　　　　　　　　　　　　　　　　　　91

The elimination of static hazards, either 0 or 1, can usually be done by adding a transition term to the control equation in addition to the prime implicants. If the term ACD (which covers or closes the transition in question) is included in the control equation, the static-1 hazard is eliminated. The resulting *hazard free* disjunctive expression is

$$Z = AB + B'CD + ACD$$

which requires seven rather than four two-input operators or, if the mixed form

$$Z = AB + CD(A + B') = AB + CD(A'B)'$$

is implemented, only five two-input operators are needed. This is certainly not optimal, as defined previously, but *may* be essential to the correct functioning of the machine.

A similar situation exists when multiple input changes occur in an automation system. Consider the system shown in Figure 6.10.

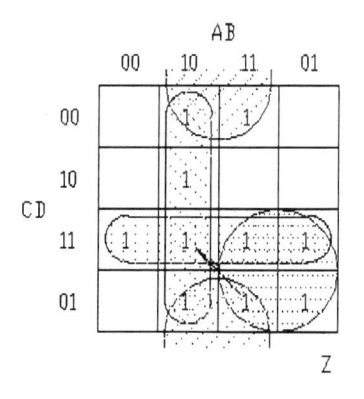

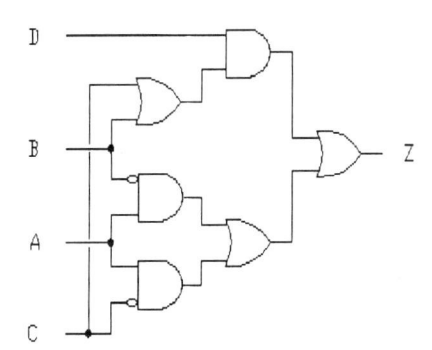

Figure 6.10a Karnaugh Map.　　　　**Figure 6.10b** Logic Implementation with Hazard.

Figure 6.10 Static-1 Logic Hazard.

This system contains a static-1 logic hazard that exists when two or more inputs are permitted to change simultaneously. Consider the situation when the present state is $ABCD = 1101$, the next state is $ABCD = 1011$, and the two state changes occur at exactly the same instant (highly unlikely, but certainly possible). Assuming that the presence of the NOT at the input of the two INHIBIT elements slows down the signal as before, it is easily seen that the output Z will be in the 0 state momentarily during the transition between states.

There are two methods that can be used to eliminate the static-1 logic hazard. One is to follow the same procedure that was used when eliminating static-1 hazards; that is, to add the required transition term. The second is to *prevent* the possibility of simultaneous switching of the variables through the use of a DELAY-IN element. The resulting systems are illustrated in Figure 6.11.

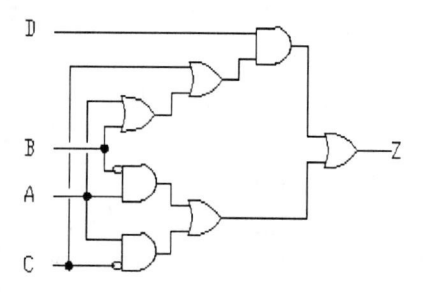

Figure 6.11a Added Closing Term.   **Figure 6.11b** Added DELAY-IN Element.

Figure 6.11 Eliminating Static-1 Logic Hazards.

A third type of hazard is the static function hazard. Consider the expression $Z = B'D' + BD$ which is shown in Figure 6.12. In the event that simultaneous changes of variable states are prohibited, there is no way in which one can reach the $ABCD = -1-1$ state from the $ABCD = -0-0$ state without creating a static-1 hazard. The only time that this hazard can be avoided is in the most favorable case, where the extremely unlikely condition of simultaneous switching occurs. This is a result of the function itself, and *cannot be corrected mathematically*.

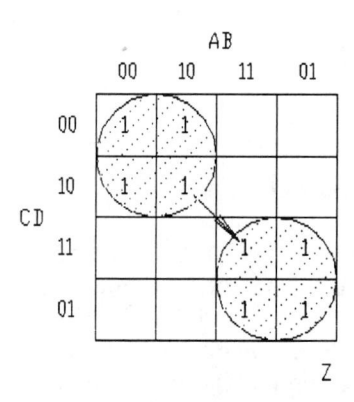

Figure 6.12 Static Function Hazard.

Note that the presence of static hazards always implies the existence of a related dynamic hazard. This condition is shown in Figure 6.13. By eliminating static hazards, the related dynamic hazards will automatically be removed.

Sec. 6.2 Typical Systems 93

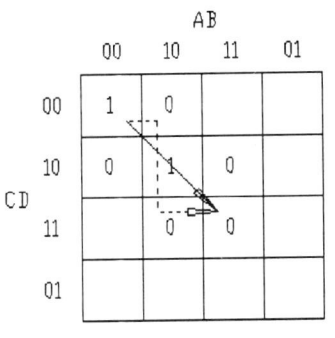

Figure 6.13 Dynamic Hazard.

While hazards can usually be eliminated mathematically, it is possible for them to be reintroduced into the control system through the selection and application of various hardware devices. Consider the OR in the relay application illustrated in Figure 6.14a when operating on the function $Z = AB + BC$. This device really develops the function

$$Z = AB + (AB)'BC$$

which can be reduced to

$$Z = AB + (A' + B')BC = AB + A'BC$$

which obviously contains a hazard as illustrated in Figure 6.14b.

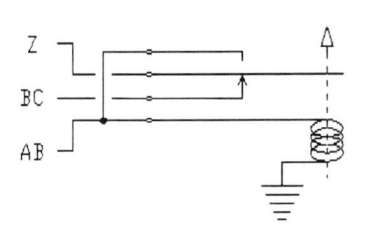

Figure 6.14a Relay OR Implementation.

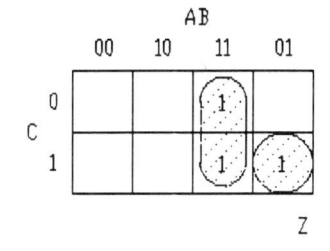

Figure 6.14b Karnaugh Map of $Z = AB + A'BC$.

Figure 6.14 Equipment Generated Static Hazard.

Just as hazards can be introduced into a system through the injudicious selection of hardware, their effects can be mitigated by using wave-shaping techniques. For example in the case of a static hazard, a momentary spurious signal can be delayed and smoothed through the use of a DELAY-IN element on the hazardous input, or a DELAY-OUT on the nonhazardous variable. The effect of this technique is to force a particular sequence in the appearance of a set of inputs at an operator.

Another hazard could arise from the unwanted repetition of an input signal. This can occur in electrical systems when contacts bounce, that is open and close repetitively in a vibratory mode. Depending on the nature of the operator being fed by such a signal, either a static or dynamic hazard could exist. In this case, the damping effect of the R-C network in the DELAY elements can eliminate the problem.

6.2.2 Hazard-Free Circuits

The most convenient way to detect potential hazards is through the use of Karnaugh maps, although in many cases the complete listing of the essential, chosen, and redundant terms developed from a prime implicant analysis is also hazard free. In order to expedite these analyses, maps and results from prime implicant analyses are shown in Figure 6.15.

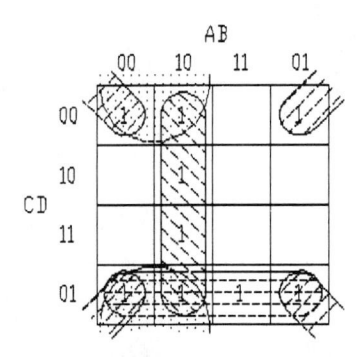

Figure 6.15a Optimum Form $Z = AB' + A'C' + C'D$; Hazard Free Form $Z = AB' + A'C' + C'D + B'C'$.

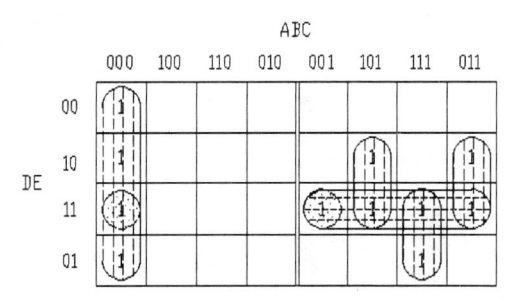

Figure 6.15b Optimum Form $Z = ABCE + AB'CD + A'BCD + A'B'C' + CDE$; Hazard Free Form $Z = ABCE + AB'CD + A'BCD + A'B'C' + CDE + A'B'DE$.

Figure 6.15 Karnaugh Maps for Hazard Determinations.

Upon analysis of both maps, it can be seen that static hazards exist in both cases. In Figure 6.15a, the transition from $ABCD = 1000$ to 0000 (and conversely, of course) shows the possibility of a static-1 hazard, as does the transition from $ABCDE = 00111$ to 00011 in Figure 6.15b. Figures 6.15c and

Sec. 6.2 Typical Systems 95

d are printouts of a microcomputer program named PRIMP. A discussion of and documentation for this program is given in a later chapter. At this point, it is enough to notice that the unused or *redundant* term, in both cases, is the one required to eliminate the hazard.

Unfortunately, dynamic hazards exist in both of the systems shown above. For example, dynamic hazards occur in the case shown in Figure 6.15a if there is a transition from $ABCD$ = 1010 to 0000 or from $ABCD$ = 1011 to 0001 if A changes state before C; or in a transition from $ABCD$ 1011 = to 1101 if B changes before C. These hazards can be easily eliminated by delaying A until after C has changed state in the first case, or by delaying B sufficiently in the second case.

Similarly, dynamic hazards exist in the case shown in Figure 6.15b when the variables ABCDE have the following state changes:

From $ABCDE$ = 00000 to 10110
00010 to 10110 or 10111 or 00111
00011 to 10110 or 10111
00001 to 00111
00111 to 10110, etc.

Of course, when changing from the second to the first state, the hazard also exists. The treatment again is to delay the appearance of one or more of the multiple inputs, which corrects the situation.

As a final example, consider the case of sorting different but similar items moving on a conveyor. This problem is encountered in packaging operations where small, medium, and large packages of the same item are processed. Figure 6.16 shows such a situation. Note that there are two different actions required: one to present a single package at a time, and a second to determine the size of the package and direct it to the appropriate location.

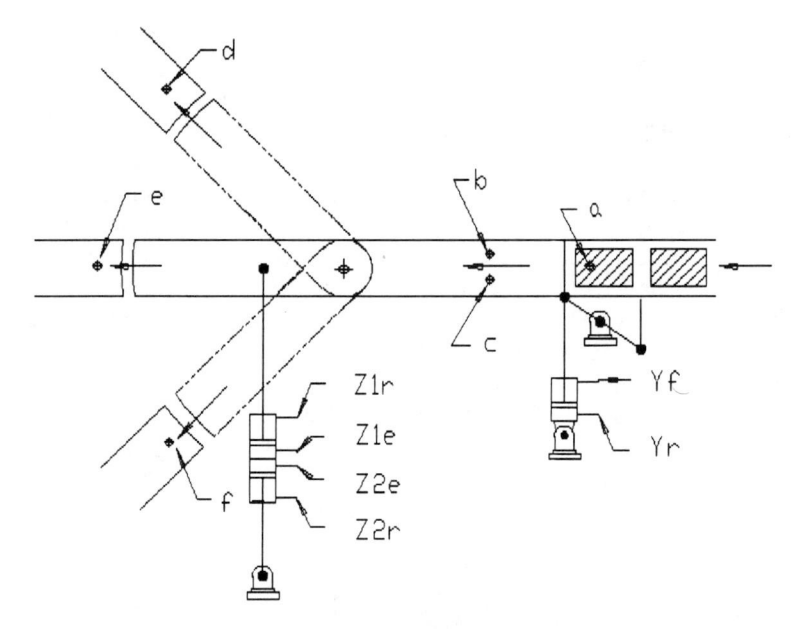

Figure 6.16 Automatic Package Sorting (Physical Layout).

Six sensors will be used to perform these operations: one to sense that a package is ready to be fed, two (in conjunction with the first) to categorize the three box sizes, and one each to detect that the carton has reached its correct destination. These sensors are marked a through f respectively.

The one-only feeder, marked Y, operates as follows. In the normal or start-up position, the cylinder is as shown. The bar or dam connected to this cylinder prevents the passage of any carton into the dispatching area. Upon receipt of either a start or a correctly dispatched signal, the output changes state, resulting in an action that blocks the second box while permitting the first box to enter the gauging area. When the package is in the gauging area, the feeder returns to its original state so it is ready to pass the next carton on demand.

Since three distinct output positions are required for this sorting, it will be necessary to shift the distributing conveyor using two power cylinders, Z_1 and Z_2, in tandem. If a SMALL carton is detected, both Z_1 and Z_2 will be retracted. If a MEDIUM carton is sensed, Z_1 will be extended and Z_2 will be retracted. If a LARGE carton is sensed, both cylinders will be extended.

The gauging logic can be described as follows:

- If no carton is ready to be fed and sorted and the feeder has been reset, then do nothing.
- If a box has been fed, and either the system has just started or the

Sec. 6.2 Typical Systems

previous box has been correctly dispatched, and neither B nor C has been actuated, then direct the package to the area marked SMALL.
- If a box has been fed, and either the system has just started or the previous box has been correctly dispatched, and only B has been activated, then direct the package to the area marked MEDIUM.
- If a box has been fed, and either the system has just started or the previous box has been correctly dispatched, and both B and C have been activated, then direct the package to the area marked LARGE.

The logic for the feeder is:

- If the system has just started or if a carton has been correctly dispatched, and if a carton is ready to be fed; then operate the feeder.
- If a part has been fed, then reset the feeder after an appropriate delay to assure that the carton has passed out of the feeder.

This completes the first phase of the design of the automatic sorting and distribution device. The operation of the system has been defined in detail through the use of a word description.

Based on the word statement of the problem, the following control equations can be written. It is suggested that the reader verify the accuracy of each equation.

$$Z1_{[RET]} = A'B'C'$$

$$Z1_{[EXT]} = A'BC' + A'BC = A'B(C + C') = A'B$$

$$Z2_{[RET]} = A'B'C' + A'BC' = A'C'(B + B') = A'C'$$

$$Z2_{[EXT]} = A'BC$$

$$Y_{[FEED]} = A(START + D + E + F)$$

$$Y_{[RESET]} = A'(DEL)$$

In addition to the system equations written above, it should be noted that certain pathological state combinations exist, for example:

- input C is ON and input B is OFF, or
- Z_1 is retracted and Z_2 is extended.

These are situations that cannot occur since there are no input combinations in our control equations that have these outputs.

There are only two viable alternatives in the event that a pathological event occurs.

- Stop the machine immediately and take corrective action, or
- Do nothing, ignore the signal, and continue the operation.

The decision as to which of these is the most appropriate is a function of the cost of being wrong. For example, the existence of the state $BC = 01$ indicates that the gauging sensors are not operating properly and that any subsequent decisions are unreliable; therefore serious consideration should be paid to finding and eliminating the source of this condition.

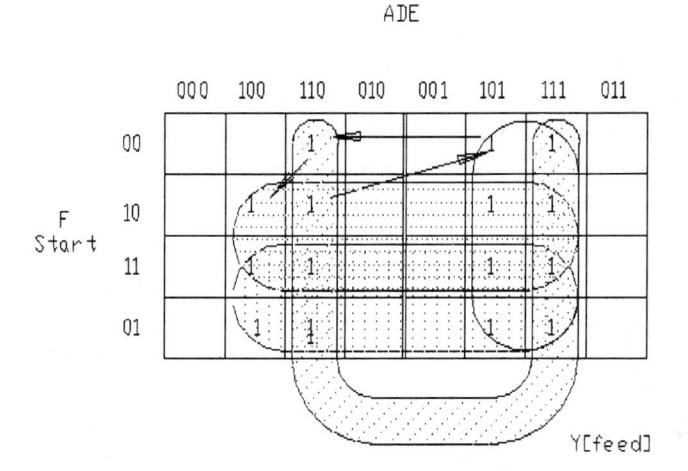

Figure 6.17 Static-1 Function Hazard Analysis.

The next design phase is the implementation in hardware and the elimination of hazards. Due to their simplistic nature, the Z expressions are hazard-free, but the $Y_{[FEED]}$ expression warrants investigation since it contains a number of ORs, a situation that is inherently hazardous. A map of this function is shown in Figure 6.17, and indicates that there are three pairs of static-1 function hazards:

From $ADEF(Start) = $ 11000 to 10010
11000 to 10100
11010 to 10100

Since this type of hazard will not respond to elimination by mathematical methods, they will have to be treated using hardware DELAYs, if they are to be treated at all! Again, the decision to treat or not is based on the cost of being wrong. Even if relatively slow AONI logic devices are used, the machine response time is *significantly* slower than the response of the input and control devices. Therefore, there should be more than enough equipment damping to eliminate the effect of the static-1 hazards.

Sec. 6.3   Problems

In this case, the decision will be made to ignore the hazards until such time as they become a problem; if they do, delays will be inserted into the appropriate lines. MPL implementation was selected based on the criteria shown in Figure 6.4. This decision would be implemented by substituting the appropriate valve symbol for the logic symbol in the multiple-input multiple-output logic system shown in Figure 6.18.

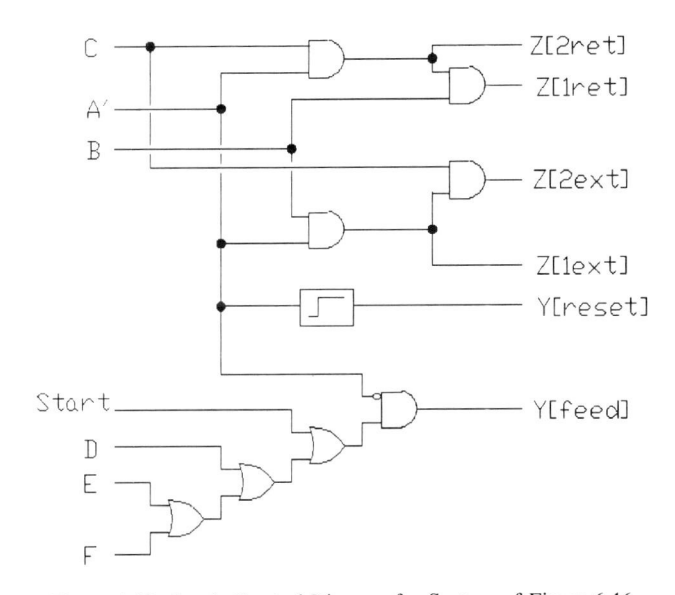

Figure 6.18   Logic Control Diagram for System of Figure 6.16.

6.3 PROBLEMS

Develop the optimal hazard-free implementation for all the following problems.

1. When operator input A and foot input B are ON, activate output $Z1$. If input C is OFF and input A is ON then energize output $Z2$. $Z2$ should also be actuated in the event that the operator is holding input A ON. There should never be an ON signal to $Z2$ in the event input B is ON. $Z1$ should be ON if C is ON.

2. A pneumatic press is to be modified such that operators are prevented from endangering themselves by placing their hands into the work area. The system requirements are as follows:
   - Two operating inputs are to be used, one each for the right and left hand.
   - The press ram can move downward if and only if both inputs are actuated. The press ram should stop immediately if either hand is removed from its input for any reason.

- The press ram can move upward if and only if both operating inputs are released by the operator.

3. A control system has two inputs, A and B, and two outputs, Y and Z. Output Y is ON whenever both inputs are ON together or are both OFF. Output Z is ON if either input is ON.

   Implement this system in the most efficient manner. Defend your selection of whatever hardware system you use.

4. A machine used for filling containers with a liquid product has two inputs and three outputs. One input indicates that the container is in the proper position for filling, while the second input indicates that the container is full. The three outputs energize the power cylinders that are actuated by means of spring-return four-way valves. The first cylinder is to move the container into the filling station. The second cylinder pushes the full container from the filling station to a conveyor. The third cylinder controls a valve that passes the material used to fill the container. It is necessary to wait for five seconds after the container is in position for filling before starting the actual filling operation. The medium used to fill the container is moderately volatile and poses some explosion hazard.

   Implement this system using the most appropriate technique. Defend your choice.

5. A machining process is to be semi-automated. There are several elements that can be used to sense various parameters. The following machine operation is desired.

   Stop the process if either dimension A or dimension B is out of tolerance and a new tool is available. If the old tool is still sharp and is adjustable, then it can be used in place of a new tool after its adjustment. If the old tool is not sharp and not adjustable, then initiate the procurement of a new tool.

   Develop the logic flow diagram for this procedure and implement it with appropriate hardware. Defend you choice of hardware.

6. Develop the logic equations and implement them using the most appropriate techniques for a coin changer. This machine will make change for dimes, quarters, and fifty-cent pieces.

   In the event that a dime is inserted, two nickels will be returned.

   If a quarter is inserted and sufficient nickels are available, one dime and three nickels will be returned, otherwise two dimes and one nickel will be delivered.

   If a fifty-cent piece is inserted, one quarter, two dimes, and one nickel are to be returned, if sufficient coins are available. In the event that two dimes cannot be delivered, one quarter, one dime, and three nickels should be dispensed. If no dimes are available, one quarter and five nickels should be returned. If an insufficient number of dimes and nickels are available to make these change combinations, two quarters are to be released.

   If it should happen that no combination of coins is possible due to a lack of change, the original coin should be returned and an annunciator energized that will display a message to that effect. The machine has weight and size sensors to determine what coin was deposited.

   In addition to the above, provision the machine to deliver one dime if two

Sec. 6.3 Problems

nickels are inserted; one quarter in the event two dimes and one nickel, one dime and three nickels, or five nickels are inserted; or a fifty-cent piece if any combination of nickels, dimes, and quarters totaling fifty cents are deposited. If a combination of coins totaling other than ten, twenty-five, or fifty cents is input, then return them and energize a display that states this.

Chapter 7

PROBLEM STATEMENTS

Before starting to solve a problem, it is always a good idea to know what problem to solve.

In the case of combinational synthesis, the problem is declared through the word statement from which the basic control expressions are derived. These can then be stated canonically through the use of truth tables, and optimized using algebraic, mapping, or prime implicant techniques.

In the case of sequential design, it is not only necessary to know the present input conditions; knowledge of the present output and the prior inputs are also required. This results in a potential dilemma when the same input state has more than one output state.

While there are many different methods used to state sequential problems, three will be discussed: the extended mapping method, the primitive flow table, and the logic specification chart.

7.1 STATEMENT OF COMBINATIONAL PROBLEMS

Combinational problems are stated as a series of input-output relationships, where the input states all occur simultaneously and each output relationship is unique. Multiple input, multiple output prime implicant techniques were discussed in Chapter 4, section 3.5. In the event that no more than six input variables are involved, a parallel and often more effective technique uses multiple output maps.

Consider the problem discussed in section 4.3.5:

$$Z_1 = a'b'c + ab'c' + ab'c + abc$$

$$Z_2 = a'b'c + a'bc' + a'bc + abc$$

$$Z_3 = a'bc' + abc' + abc$$

The map for this multiple output system is shown in Figure 7.1.

C \ A B	0 0	1 0	1 1	0 1
0	0 0 0	1 0 0	0 0 1	0 1 1
1	1 1 0	1 0 0	1 1 1	0 1 0

Z_1 Z_2 Z_3

Figure 7.1 Multiple Output Map.

It is evident from a quick perusal of the map that the input *abc* is shared by all three outputs, and that *a'bc'* and *a'b'c* are shared by two. Unfortunately, the use of these combinations will not result in a hazard-free system.

Since we want to implement hazard free systems *unless we have specific reason not to*, the following control equations were selected:

$$Z_{1hf} = ab' + b'c + ac$$
$$Z_{2hf} = a'b + a'c + bc$$
$$Z_{3hf} = ab + bc'$$

If "a", "b", and "c" are used as the input literals and an AONI hardware system is used, these equations can be implemented with five INHIBIT, three AND, and five OR elements resulting in a score or index of merit of $3 + 13 = 16$. Considering the control equations developed in Figure 4.6, implementation requires five INHIBIT, two AND, and five OR elements; for an index of 15; which is optimal, but not hazard free.

In all fairness, it should be noted that the hazard free system can be derived from the prime implicant analysis:

$$Z_1 = ab' + a'b'c + abc = ab'(c + c') + a'b'c + abc$$
$$= ab'(c + c') + b'c(a + a') + ac(b + b') = ab' + b'c + ac$$
$$Z_2 = a'b + a'b'c + abc = a'b(c + c') + a'c(b + b') + bc(a + a')$$
$$= a'b + a'c + bc$$
$$Z_3 = bc' + abc = bc'(a + a') + abc = bc' + ab(c + c') = bc' + ab$$

Which is all well and good, but it presupposes that the existence of hazards, and the appropriate corrective actions, can be readily determined from the prime implicant table. This is not always the case.

In summary, the most optimal hazard free statement of a combinational problem can usually be obtained from a map if six or fewer input variables are present. In the event that more inputs exist, the prime implicant technique, coupled with some intuitive knowledge and algebraic reductions, is the most productive design procedure.

7.2 STATEMENT OF SEQUENTIAL PROBLEMS: EXTENDED MAPPING TECHNIQUE

One of the advantages of the mapping technique is that it assists in the visualization of a problem by graphically representing the problem statement.

Consider a two cylinder system where the desired pattern of operation is:

- Extend cylinder 1, then

- Extend, then retract cylinder 2, then
- Retract cylinder 1

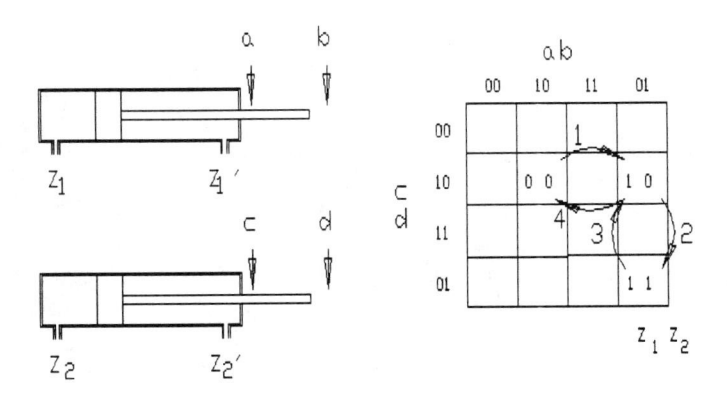

Figure 7.2 Two Cylinder Sequencing Problem.

This system is illustrated in Figure 7.2. The four output steps that will be sequenced have the following system equations:

$$Z_1' = ab' \qquad Z_1 = a'b$$
$$Z_2' = cd' \qquad Z_2 = c'd$$

Assume that the start-up position of the system is Z_1' and Z_2'. Then the sequencing for a complete single cycle should be:

$$Z_1'Z_2' \rightarrow Z_1Z_2' \rightarrow Z_1Z_2 \rightarrow Z_1Z_2' \rightarrow Z_1'Z_2'$$

This is an ongoing IF-THEN, or IMPLICATION relationship of classical logic. While there are no hardware devices that create IMPLICATION, it is defined by the truth table shown below in Figure 7.3.

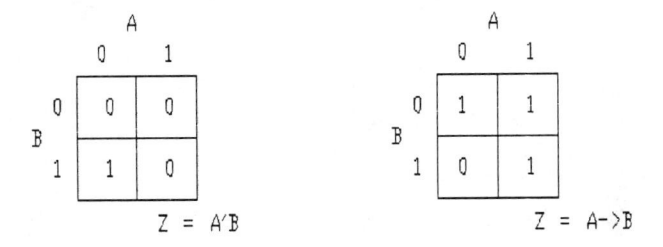

$Z = A'B$ 　　 $Z = A \rightarrow B$

Figure 7.3 Logic Comparison INHIBIT/IMPLICATION.

Particular attention should be placed on the fact that IMPLICATION is NOT INHIBIT. This operator has been largely ignored by control and

Sec. 7.2 Statement of Sequential Problems

automation designers; it will be more fully discussed later. In terms of IMPLICATION, the above sequence of relationships can be written:

$$\text{If } \{Z_1'Z_2'\} \text{ then } \{Z_1Z_2'\} \qquad \text{or} \qquad (Z_1'Z_2') \to (Z_1Z_2')$$

which is designated by the unidirectional arrow. Note that each arrow expresses a unique relationship; there are four of these.

In this simple system, an investigation of the sequence shows that position Z_1Z_2' has two possible outcomes, or *next states* that are not differentiated by the existing logic. The map of the system in Figure 7.2 shows this with great clarity.

The obvious solution to this problem is to discriminate through the introduction of another variable. This variable must *remember* the state prior to Z_1Z_2', and will be called Y. The system equations are now:

$$Z_1'Z_2' = ab'cd' \qquad Z_1Z_2' = a'bcd' \qquad Z_1Z_2 = a'bc'd$$

$$\text{Set } Y = Y_S = Z_1'Z_2' = ab'cd' \qquad \text{Reset } Y = Y_R = Z_1Z_2 = a'bc'd$$

and the sequence becomes:

$$Y_S = Z_1'Z_2'$$

$$Z_1'Z_2' \to Z_1Z_2'$$

$$Z_1Z_2' \to Z_1Z_2$$

$$Z_1Z_2 \to Y_R$$

$$Y_R = Z_1Z_2'$$

$$Z_1Z_2' \to Y_S$$

So far, all the relationships have been expressed in canonical form, which, while certainly correct, is far from optimal. The expressions can be minimized in the following manner:

- Analyze the inputs for combinations that cannot physically exist and eliminate these from consideration when optimizing.
- Consider all canonical terms that have not been expressly used as having no effect on the system and therefore falling in the category of *don't cares*.
- Combine the cells in the map so as to produce a near optimal hazard free system that changes the state of the memory(ies) at the last possible moment.

The results, using the numbering system of Figure 7.2 and the above steps, are shown in Figure 7.4.

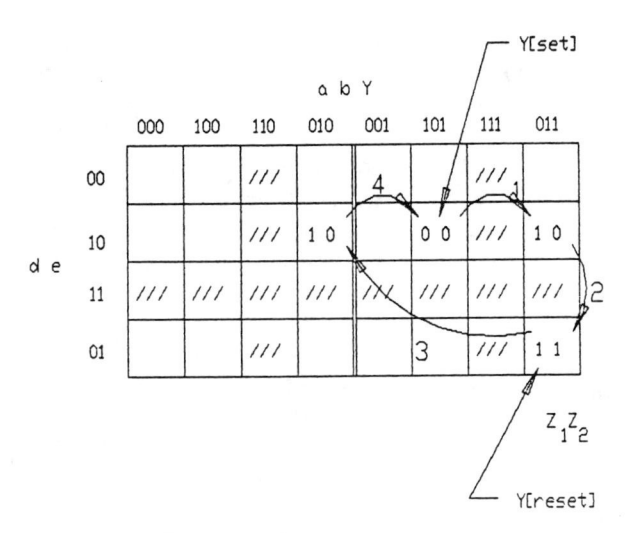

Figure 7.4 System Optimization.

From this map, the system equations can be developed and are:

$\{Y_S = Z_1'Z_2' = a\} \rightarrow Z_1Z_2'$ $\{Z_1Z_2' = bY\} \rightarrow Z_1Z_2$

$\{Y_R = Z_1Z_2 = d\} \rightarrow Z_1Z_2'$ $\{Z_1Z_2' = cY\} \rightarrow Z_1'Z_2'$

It should be noted that while the IF-THEN procedure might appear a bit casual, it is in reality, as stated earlier, very rigorous and has been defined using Boolean terms. Also, note that the number of memories required increase by a power of two relationship: one memory will differentiate between two confounded paths, two memories among three or four paths, three memories among five to eight paths, and so on. The system designed in the preceding procedure is shown in Figure 7.5.

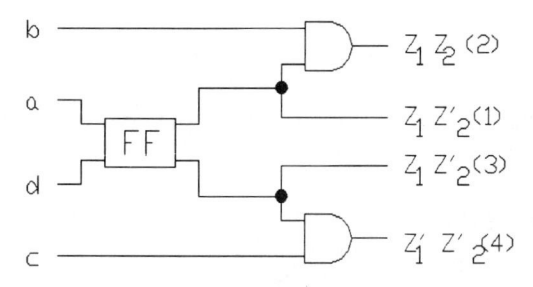

Figure 7.5 Logical System Design.

The extended mapping technique is much more than a method of stating a control or automation problem. In reality it is a full design technique that is insidiously attractive due to its simplicity and clarity. It has been presented here in its entirety since anything else would be counterproductive.

Unfortunately, there are some significant limitations to its use. It is effective only for deterministic problems, that is problems where there is a distinct and repetitive pattern in the sequence of operation of both the inputs and the outputs of an automation system. Also, as in combinational problems, the use of a map with more than six variables is extremely difficult. Within these constraints, the technique is valuable.

7.3 STATEMENT OF SEQUENTIAL PROBLEMS: PRIMITIVE FLOW TABLES

While there are many deterministic control systems, especially in the automation of production equipment, there are control and automation systems that are not regular and repetitive. These systems are called stochastic or probabilistic and are characterized by their lack of regularity in the appearance of inputs and combinations of inputs. Sorting and inspection systems, distribution systems, or any random input system fall into this category.

The IF-THEN technique presented in the previous section is generally unusable in the analysis of stochastic systems due to the complexity of the relationships and the number of statements required. For example, with only three inputs, there are eight possible input combinations.

Shortly after Karnaugh published the mapping method for combinational logic synthesis, D. A. Huffman[19] developed a technique called the *Primitive Flow Table* or PFT, to describe sequential systems; both deterministic and stochastic. This scheme provides a *record* of input combinations and the present output state, as well as the next output states. It is an effective way to *document* a history of inputs and outputs.

A PFT for a system with n input variables consists of 2^n input columns; one for each possible input combination arranged by Gray Code. An additional column on the extreme right shows the output combinations. Each row in the table shows a *stable* machine state, that is one that has a distinct output combination associated with it, the next possible unstable machine states, and the output associated with the stable state.

A *stable* state also refers to the condition where no change of the state of the machine can occur without some change in the input state. An *unstable* state is a transition state that indicates the next stable state should the appropriate change in the input state occur.

The PFT for the system discussed in section 7.2 is shown in Figure 7.6. Note that the starting point of the PFT is purely arbitrary, but is normally taken as the starting or *turn-on* state. Also note that the stable states are indicated by square brackets ([ ]).

	NEXT STATES																OUTPUT STATES	
	INPUT STATES: In																	
n	1	2	3	4	5	6	7	8	9	10	11	12	13	14	15	16	Z1	Z2
a	0	1	1	0	0	1	1	0	0	1	1	0	0	1	1	0		
b	0	0	1	1	1	1	0	0	0	0	1	1	1	0	0			
c	0	0	0	0	1	1	1	1	1	1	1	1	0	0	0	0		
d	0	0	0	0	0	0	0	0	1	1	1	1	1	1	1	1		
		2			[1]												0	0
		[2]										3					1	0
				4								[3]					1	1
		[4]			1												1	0

Figure 7.6 Primitive Flow Table.

Three things can be seen from the above example:

- The elegant simplicity of the technique in charting the system flow,
- The ease of identifying the confounded states by recognizing the existence of more than one stable state in any one column, and
- The waste of space and effort required in establishing columns for each possible input combination in a deterministic system.

The rules for creating a PFT are:

- Determine the number of variables, therefore the number of columns required in the table,
- Code the inputs by Gray Code and create the table,
- *Select* the initial condition and name that input state stable state [1]; specify the output state,
- Determine the unstable next-states by permitting the change of one and only one variable at a time; record these states, and
- Select the next stable state from one of the previous unstable states, repeat the process.

Consider the case of three fair coins, *a*, *b*, and *c*. These coins will be flipped, *one at a time*. All the coins are *selected* to be heads initially. It is desired that ultimately they all be tails. Figure 7.7 illustrates the PFT of this stochastic system. The coding for the outputs should be evident to the reader.

Sec. 7.4 Statement of Sequential Problems

	NEXT STATES								OUTPUT STATES
	INPUT STATES: In								
n	1	2	3	4	5	6	7	8	Z1 Z2 Z3
a	0	1	1	0	0	1	1	0	
b	0	0	1	1	1	1	0	0	
c	0	0	0	0	1	1	1	1	
	[1]	2	3					4	H H H
	1	[2]	5			6			T H H
	1		5	[3]	7				H T H
	1				7		6	[4]	H H T
		2	[5]	3		8			T T H
		2				8	[6]	4	T H T
				3	[7]	8		4	H T T
						[8]			T T T

Figure 7.7 Primitive Flow Table.

While it is not recommended that the PFT be used in this trivial manner, it certainly illustrates the advantages of the technique. One may start at stable state [1] and proceed through state [2], state [5], or state [6], and then to [8]; or from state [1] to state [3], then to state [5] or state [7], and then to state [8], and so on. There are no possible paths that cannot be traced from [1] to [8]. It is for this reason that the statement was made to the effect that a complete system history is generated.

7.4 STATEMENT OF SEQUENTIAL PROBLEMS: LOGIC SPECIFICATION CHARTS

It has been shown that for small deterministic problems, the extended mapping technique is effective. For stochastic problems, the Primitive Flow Table is useful. Both techniques have significant shortcomings that render them ineffective for the general analysis and synthesis of systems.

To be truly effective, a method must not only be able to describe large-scale problems, but it must also be adaptable to use on any computer but preferably a microcomputer. Such a method is the *Logic Specification Chart* or LSC.

The LSC shares the good features of the PFT, without the one drawback of having to specify all the unused input combinations. The LSC is constructed in a manner similar to the PFT with the following exceptions:

- The output states are shown in leftmost columns.
- Columns are assigned only to the input states that are used, and then only when required. Gray coding is not necessary.

The rules for creating a LSC are similar to those used with the PFT:

- *Select* the initial condition. Identify the input state and record the values of the input variables in the first input column. Name the stable state [1] and record it in the first row, first column.
- Assign the output state for the first input combination and record it in the first row of the output column.
- Determine *all* of the next possible states by permitting the change of one and only one variable at a time. Record these next states in the columns to the immediate right of, and in the same row as the stable state. If an input combination already exists, *do not assign a new input state number*; use the old one.
- Repeat the above steps until all the possible states of the system have been duly noted and recorded.

Figure 7.8 is the first example of the LSC and shows the two-cylinder system described in section 7.2 as illustrated in Figures 7.2 and 7.6. The similarity to the PFT in Figure 7.6 is unmistakable, as is the LSC's efficiency and economy.

OUTPUT STATES	NEXT STATES		
Z1 Z2	INPUT STATE: In		
	n	1	2	3
	a	1	0	0
	b	0	1	1
	c	1	1	0
	d	0	0	1
0 0		[1]	2	
1 0			[2]	3
1 1				4	[3]
1 0			1	[4]

Figure 7.8 Logic Specification Chart.

Sec. 7.4 Statement of Sequential Problems

In addition to the purely logical considerations, there are several significant details that must be considered in addition to the basic logic requirements when designing real systems:

- How will the system be started and restarted or reset after conditions such as power failures or system breakdown?
- How will components other than the logic elements be incorporated into the system to satisfy the design requirements?
- What additional techniques can be used to effect system optimization including the appropriate use of don't-care combinations?

When considering items such as these, one must balance the alternatives against the do nothing option. Doing nothing is always viable as long as it is a *considered* decision, not one which is arrived at by default.

How often can power failures be expected? What action should be taken by either the machine or its operator during or after such a failure? Should the machine be started and restarted automatically or manually? Should the machine return to its initial condition or continue in its cycle in the event of a failure? Should the machine be made foolproof?

All of these questions (and many others) are significant and should be considered under three criteria: technical feasibility, economic desirability, and fiscal reality. All three of these criteria have been discussed in the first chapter. In the event that they were not clear at that time, it is strongly suggested that this material be reread now.

As additional features are incorporated into the system, so are complexities that invariably affect cost and reliability. Just as it is basic that one does not fix something that is not broken, one does not increase complexity unless there are good functional and economic justifications.

Consider for example, the two-cylinder system described in Figure 7.8 with automatic reset to the start-up condition in the event of power failure. A power ON/OFF sensor will be required and will be designated e. In the event of failure, the machine will have to be manually restarted to provide for fault tracing and repair. The LSC for this system is shown in Figure 7.9. Upon comparing the LSCs in Figures 7.8 and 7.9, the increase in the complexity of the system is obvious.

OUTPUT STATES $Z_1 Z_2$	NEXT STATES INPUT STATE: In
	n 1 2 3 4 5 6 7 8 9 10 11
	a 0 1 0 0 1 0 1 0 0 0 0
	b 0 0 1 1 0 0 0 1 1 1 0
	c 0 1 1 0 1 1 0 1 0 0 0
	d 0 0 0 1 0 0 0 0 0 1 1
	e 0 1 1 1 0 0 0 0 0 0 0
0 0	[1] 2
0 0	[2] 3 6
1 0	[3] 4 9
1 0	5 [4] 11
1 1	2 [5]
0 0	[6] 7 8
0 0	1 [7]
0 0	1 [8]
1 0	7 [9] 10
0 0	1 [10]
1 1	10 [11] 12
1 1	1 [12]

Figure 7.9 Logic Specification Chart with Power Failure Provision.

There are many cases when signal treatment is required. This has been demonstrated in combinational systems and is at least equally as valid when considering sequential controls. As in the earlier case, DELAY-IN, DELAY-OUT, and ONE-SHOT elements are the principal non-logic devices used to satisfy system requirements.

As in any system, optimization is always desirable. However, hardware minimization does not necessarily imply project optimization. The cost of

7.5 SIMPLIFICATION OF LOGIC SPECIFICATION CHARTS

The purpose of the LSC is to provide a history of the sequential changes in the input and output states of a control system. As such, it is essential that when this record is created, it is *absolutely* accurate. Economy of presentation and elimination of redundancies are *not* goals when the initial draft is made, nor should they be. It is far better to include duplicates than to omit one required state.

Once an accurate draft is available, a search can be undertaken to determine if two or more stable states are equivalent to each other. Stable states are equivalent iff (if and only if):

- The states under consideration have the same input combination; that is, they lie in the same column of the LSC,
- The outputs for the states under consideration are either identical or contain don't-care entries so that at the least they are equivalent, and
- The next state entries for the states under consideration are either equivalent (note: *not* identical) or contain don't-care entries so that at the least they are equivalent.

Consider the LSC shown in Figure 7.10

OUTPUT STATES		NEXT STATES			
$Z_1 Z_2$		\multicolumn{4}{c	}{INPUT STATE: In}
$Z_1 Z_2$	n:	1	2	3	4
	a	0	1	1	0
	b	0	0	1	1
0 0		[1]	3	4	2
- 1		1	-	8	[2]
0 1		6	[3]	5	7
1 1		-	9	[4]	2
1 -		-	3	[5]	12
0 0		[6]	9	5	11
1 1		1	-	14	[7]
1 -		6	9	[8]	2
0 1		1	[9]	10	2
1 0		-	3	[10]	2
0 1		1	-	4	[11]
0 1		6	-	13	[12]
1 -		-	-	[13]	7
1 0		-	9	[14]	16
- 1		-	-	[15]	2
0 -		6	3	15	[16]

Figure 7.10 Logic Specification Chart.

On investigation of the stable states in the first column, it can be said that states [1] and [6] are equivalent iff their outputs are the same (they are), state [3] is equivalent to state [9], state [4] to state [5], and state [2] to state [11]. It is readily seen that an analysis of this nature could be a monumental task in the event an orderly procedure is not followed. The use of the *Equivalent Pairs Chart* or EPC, is such a procedure.

The charts are constructed on a column by column basis. For the LSC under study, the charts are shown in Figure 7.11.

Sec. 7.5   Simpification of Logic Specification Charts

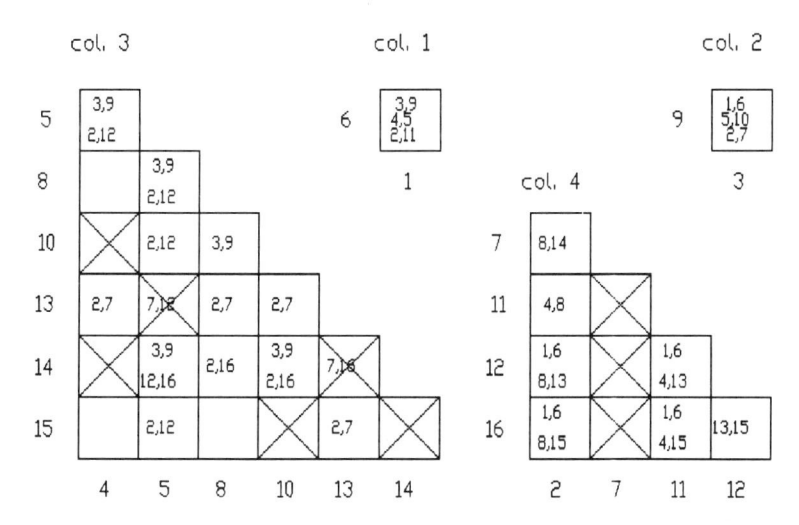

col. 3

	4	5	8	10	13	14
5	3,9 / 2,12					
8		3,9 / 2,12				
10	✗	2,12	3,9			
13	2,7	7,✗	2,7	2,7		
14	✗	3,9 / 12,16	2,16	3,9 / 2,16	7,✗	
15	✗	2,12		✗	2,7	✗

col. 1

	6	1	7	11	12	16
6	3,9 / 4,5 / 2,11					

col. 4

	2	7	11	12
7	8,14			
11	4,8	✗		
12	1,6 / 8,13	✗	1,6 / 4,13	
16	1,6 / 8,15	✗	1,6 / 4,15	13,15

col. 2

	9	3
9	1,6 / 5,10 / 2,7	

Figure 7.11   Equivalent Pairs Chart. Data from Figure 7.10.

The technique for creating or reading the EPC is simple. The coordinates of each cell are the states that are being checked for equivalency. The contents of each cell are the *next* states that must be equivalent.

The EPC for column 1 has been discussed above. Column 2 has only two stable states, [3] and [9]; their outputs are the same (01). States [1] and [6], [5] and [10], and [2] and [7] must be compared for equivalency.

The numbering scheme for the EPC of column 3 should be noted. The lowest numbered stable state is eliminated from the rows, and the highest numbered stable state is eliminated from the column arrangement. In this way, all combinations are listed with no redundancies.

In column 3, states [4] and [10] are not equivalent since their outputs differ, therefore an X is placed in that box. Similarly, states [4] and [14], [10] and [15], and [14] and [15] are eliminated, as are states [7] and [11], [7] and [12], and [7] and [16] in column 4.

On the elimination of states [7] and [12], it is noticed that the equivalency of this pair is required at states [5] and [13]; therefore the latter pair is eliminated at this time. State pair [13] and [14] is eliminated since states [7] and [16] are not equivalent.

Now that the equivalency of the stable states has been established, they should be combined so that a new *Reduced Specification Chart* or RSC can be developed. The RSC contains the minimum number of stable states that can be used to describe the system. This process is accomplished through the use of a merging graph and state tables in a manner similar to that used with Prime Implicant Tables.

The premises on which the merging graph is constructed are:

- Two equivalent stable states can be grouped together to form a single new state, and
- More than two stable states are equivalent and therefore may be grouped iff each and every possible combination of two states are equivalent.

The merging graph for the EPC of Figure 7.11 is shown in Figure 7.12. Note that this graph is generated by establishing nodes for each stable state being considered for equivalency, and then connecting the nodes iff the states are in fact equivalent. The merging is evidenced by the creation of complete polygons, where each node is connected to every other node in the figure. Each input state, or column, has its own merging diagram. Those for states [1] and [6], and [3] and [9] are shown but are trivially obvious. The same cannot be said for columns 3 and 4 of the LSC. For purposes of clarity, consider the graph for column 4 first.

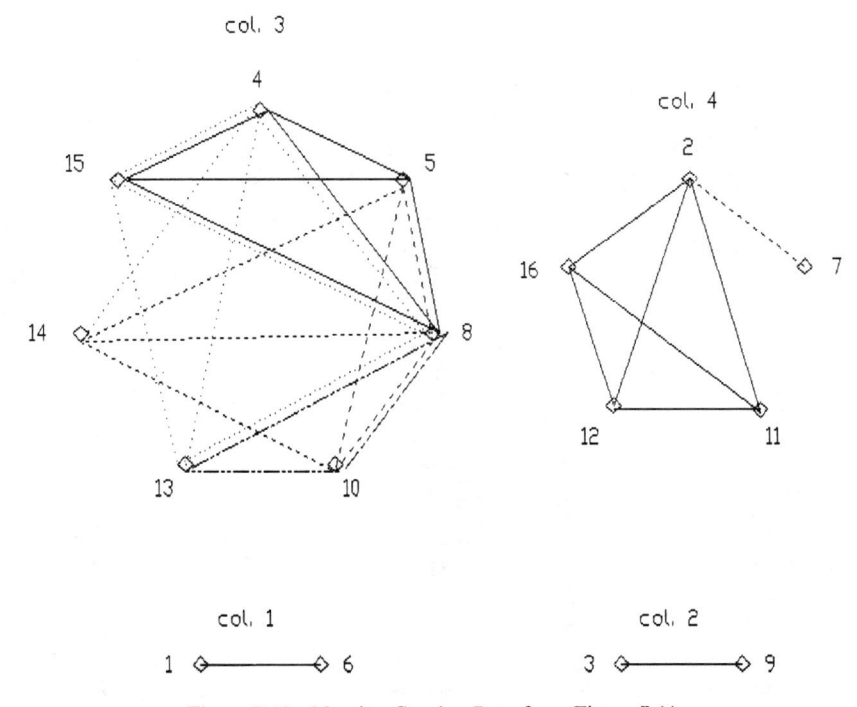

Figure 7.12 Merging Graphs. Data from Figure 7.11.

Once the equivalent pairs are noted by connecting the appropriate nodes, it is evident that nodes 2, 11, 12, and 16 form a complete polygon as defined above. Therefore they are all equivalent and can be combined into

Sec. 7.5 Simpification of Logic Specification Charts   119

a new state. Note that state 2 is also equivalent to state 7 and they combine to form another new state.

The merging graph for column 3 is much more complex. A careful analysis shows that states 4, 5, 8, and 15; 4, 8, 13, and 15; 5, 8, 10, and 14; and 8, 10, and 13 form four new combined stable states. The recording of these combined or *maximal* sets, and the determination of the essential sets are illustrated in the State Table shown in Figure 7.13.

MAXIMAL SETS	STATES																
	1	2	3	4	5	6	7	8	9	10	11	12	13	14	15	16
* (1,6)	[X]					[X]										
* (2,7)		(X)					[X]									
* (3,9)			[X]						[X]							
A (4,5,8,15)				X	(X)			(X)							X	
B (4,8,13,15)				X				(X)					X		X	
* (5,8,10,14)					(X)			(X)		(X)				[X]		
C (8,10,13)								(X)		(X)			X			
*(2,11,12,16)		(X)									[X]	[X]				[X]

Figure 7.13 State Table. Data from Figure 7.12.

This table is constructed in the same manner as the Prime Implicant Table described in Section 4.3.3. For those who feel it necessary, that section should be reviewed at this time. Since it becomes apparent from an analysis of this table that states 4, 13, and 15 are not covered by the essential maximal sets, at least one supplemental set is required. The selection of the supplemental set is made by inspection of the unbracketed Xs, or by an algebraic analysis. If the nonessential states are named A, B, and C respectively, then a conjunctive logic expression can be drawn by combining the rows containing the unbracketed Xs. For example, this table could be expressed as:

$$Z = (A + B)(A + B)(B + C) = AB + AC + B + BC = B + AC$$

The *minimum* term that will result in Z is B, therefore this state should be selected as the supplemental state. This is the same conclusion as was reached before by inspection.

The results of this analysis are entered onto a Next State Table illustrated in Figure 7.14. Note that line 7 (or new state 7) has been added to the table since the next state (4, 5) was not available without it. Note also that the missing set was underlined and the *new* set checked to identify where the problem lay. The necessity for this new state was discovered from a state-by-state analysis to make sure that, for every next state required, a selected state appears. Also notice that the states have been renumbered.

The new numbers merely *name* the combined old sets. Care should be taken not to confuse the new names of the sets with their old names.

NO.	SELECTED SET	NEXT STATE
1	(1,6)	(3,9)(4,5)(2,11)
2	(2,7)	(1)( )(8,14)
3	(3,9)	(1,6)(5,10)(2,7)
4	(2,11,12,16)	(1,6)(3,9)(4,8,13,15)
5	(4,8,13,15)	(6)(9)(2,7)
6	(5,8,10,14)	(6)(3,9)(2,12,16)
7	(4,5,8,15)	(6)(3,9)(2,12)

Figure 7.14 Next State Table. Data from Figures 7.10 and 7.13.

The selected sets are determined from the essential sets in the state table. The next states are determined by scanning the appropriate rows in the LSC. For example, state number 5 (4, 8, 13, and 15) has only next state 6 in column 1, next state 9 in column 2, and both states 2 and 7 in column 4.

The Reduced Specification Chart can be constructed using the new state numbers from Figure 7.14, and the original input and output states. This results in the most efficient statement of the automation problem. The RSC for this system is shown in Figure 7.15. Note that the state numbering in the RSC is drawn in toto from the Next State Table. For example, new state 1 was formed from old (1, 6), new state 4 from old (2, 11, 12, and 16), old next state (2,11) is contained in new state 4, and so on.

OUTPUT STATES		NEXT STATES			
		INPUT STATE: In			
$Z_1 Z_2$	n:	1	2	3	4
	a	0	1	1	0
	b	0	0	1	1
0 0		[1]	3	7	4
1 1		1	–	6	[2]
0 1		1	[3]	6	2
0 1		1	3	5	[4]
1 1		1	3	[5]	2
1 0		1	3	[6]	4
1 1		1	3	[7]	4

Figure 7.15 Reduced Specification Chart.

7.6 PROBLEMS

1. Derive the PFT that describes the system in Chapter 6, problem 6.2.
2. Analyze and design an optimal control system for the container-filling machine described in Chapter 6, problem 6.4. Use the extended mapping technique.
3. Construct a Primitive Flow Table for the coin changer problem in Chapter 6, problem 6.6.
4. Describe the desired sequence of operation for the machine process of problem 6.5, Chapter 6. Use the LSC in your answer. Simplify the LSC to its corresponding RSC, if possible.
5. On a television quiz show, three contestants are asked a question simultaneously. A contestant wishing to answer operates a button on their desk. Whoever operates their button first is permitted to answer the question.

    A logic circuit is desired that will determine which contestant should have the opportunity to answer the question. This is to be indicated by a light turning ON at the desk of the first contestant to respond. The remaining lights are to be held OFF.

    Assume that no two events can occur simultaneously and that changes can occur only one-at-a-time. The show host has a reset button; so provisions for automatic reset are unnecessary.

    Define the inputs and outputs for the system, derive a PFT for the sequential system, and develop the RSC.

    Consider the three cylinders shown below. In the four following problems, let E represent an EXTENDED piston, and R represent a RETRACTED piston. Let the piston positions be the input states and the flow inputs to the cylinders be the output states. Develop the LSCs and the RSCs (when possible) for each of the following patterns of operation.

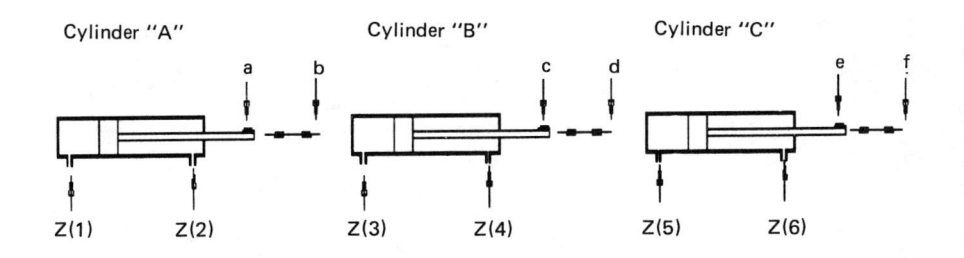

Cylinder "A" a b
Cylinder "B" c d
Cylinder "C" e f

Z(1) Z(2) Z(3) Z(4) Z(5) Z(6)

Problem 6

Cyl A: R E E R E E R R
Cyl B: R R E E R E E E
Cyl C: E R R E R R R E

Problem 7

Cyl A: R E R E R E E E R
Cyl B: R E E R R R E R R
Cyl C: R R E E R R R E E

Problem 8

Cyl A: R E R E R R E R
Cyl B: R R E E E R E E
Cyl C: E R R R E E E E

Problem 9

Cyl A: R E R R E E R R E
Cyl B: R R R R R E R R R
Cyl C: E E R E E E E R E

Sec. 7.6   Problems

For each of the following problems, derive the RSC.

Problem 7.10

OUTPUT STATES		NEXT STATES				
		INPUT STATE: I_n				
Z1	Z2	h	1	2	3	4
		a	0	0	1	1
		b	0	1	1	0
0	0		[1]	2	3	
0	1		1	[2]		4
1	0		1		[3]	5
1	1			6	8	[4]
1	1			2	3	[5]
0	1		7	[6]		4
1	1		[7]	6	8	
1	0		9		[8]	5
0	0		[9]	2	8	

Problem 7.11

OUTPUT STATES		NEXT STATES				
		INPUT STATE: I_n				
Z1	Z2	h	1	2	3	4
		a	0	0	1	1
		b	0	1	1	0
0	0		[1]	2	3	4
0	1		5	[2]	6	7
1	1			8	[3]	4
–	1		1		6	[4]
0	0		[5]	8	6	10
1	–			2	[6]	10
1	1		1		9	[7]
0	1		1	[8]	9	4
1	0			2	[9]	4
0	1		5		3	[10]

Problem 7.12

OUTPUT STATES		NEXT STATES				
		INPUT STATE: I_n				
Z1	Z2	h	1	2	3	4
		a	0	0	1	1
		b	0	1	1	0
0	1		[1]	2		3
0	–		4	[2]	5	
–	0		6		7	[3]
0	1		[4]	8		9
1	–			2	[5]	3
1	0		[6]	2		10
1	1			8	[7]	10
0	1		1	[8]	5	
1	0		6		7	[9]
0	0		11		12	[10]
1	0		[11]	8		3
1	0			13	[12]	9
0	0		4	[13]	5	

Chapter 8

SEQUENTIAL LOGIC SYSTEMS: COUNTING AND STEPPING METHODS

The essential logic of any sequential control system, regardless of whether it is deterministic or stochastic, can be simply stated. First one establishes the *current* steady state. Then one determines the appropriate *next* steady state. Finally, one develops a scheme to assure a reasonable transition between the two.

The simile between this process and the use of a road-map should not be lost. When using a road-map, one first determines where one is. Then the destination point is selected. Then and only then does one find the *best* route.

Up to this point, it has been suggested that the appropriate design technique for sequential systems is to identify input combinations that result in *confounded* outputs and to discriminate among them through the use of MEMORY devices.

It will now be demonstrated that there are are other schemes which, although usually not resulting in optimal control systems, do present viable solutions to control problems due to their directness, simplicity, and ease of implementation. These latter techniques depend to a great extent on the attributes of specific pieces of hardware.

8.1 THEORETICAL BASIS

Consider the system illustrated in Figures 7.2 and 7.8 and repeated below in Figure 8.1

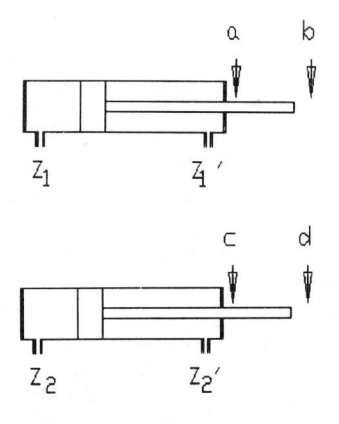

Figure 8.1a Functional Diagram. 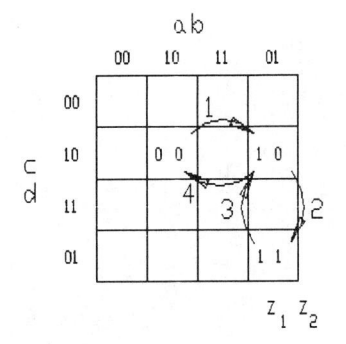   **Figure 8.1b** Karnaugh Map.

Figure 8.1 Two Cylinder Sequencing Problem.

Sec. 8.1   Theoretical Basis   **127**

OUTPUT STATES	NEXT STATES			
Z1 Z2	INPUT STATE: In			
	n	1	2	3
	a	1	0	0
	b	0	1	1
	c	1	1	0
	d	0	0	1
0 0		[1]	2	
1 0			[2]	3
1 1			4	[3]
1 0		1	[4]	

Figure 8.1c Logic Specification Chart.

This problem can be expressed by the two following sets of statements.

- If one is in state [1] AND the input state $= I_1 = ab'cd'$ then go to state [2]. If one is in state [2] AND the input state $= I_2 = a'bcd'$ then go to state [3]. If one is in state [3] AND the input state $= I_3 = a'bc'd$ then go to state [4]. If one is in state [4] AND the input state $= I_2 = a'bcd'$ then go to state [1] . . .
- If one is in state [1] AND the input state is NOT I_1 then stay in state [1]. If one is in state [2] AND the input state is NOT I_2 then stay in state [2]. If one is in state [3] AND the input state is NOT I_3 then stay in state [3]. If one is in state [4] AND the input state is NOT I_2 then stay in state [4] . . .

Notice that there are *no* confounded states. Each statement is truly independent of the others since a unique condition exists, namely the appearance of a specific and different present state in each statement.

Note also that each of the above logic statements can be reduced to an optimal combinational form prior to its implementation in the final AND statement. This does not imply that the overall system is *logically* optimal. However, when one considers the economic and fiscal constraints imposed by design and maintenance requirements, the final system developed in this manner could be the most desirable.

In the above problem, consider the combinational statement relating to state [2]. If one is in state [1] AND input state $= I_1$, OR if one is IN STATE [2] AND the input states is NOT I_2; then go to (stay in) state [2]. A similar set of statements could be made for state [4]. The set of expressions that models this system is then:

$$[2] = \{[1]I_1 + [2]I_2'\} \rightarrow [2] = [1]I_1 + [2]I_2' + [2]'$$
$$= [1]I_1 + I_2' + [2]' = [1]I_1 + ([2]I_2)'$$

On referring to Figure 3.6, it can be seen that this expression is of the same form as the Type Y Memory. It can be further reduced to

$$[2] = (([1]I_1)'([2]I_2))'$$

Which is implemented as shown in Figure 8.2.

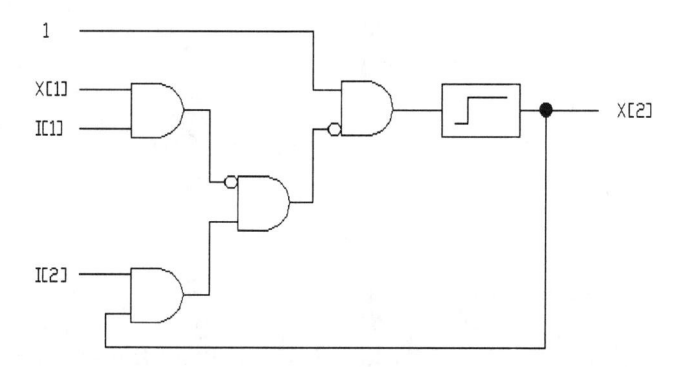

Figure 8.2 Logical Implementation of $X\,2 = ((X[1]I[1])'(X[2]I[2]))'$.

The problem is to delay the turning ON of the state [2] signal until all other output states are OFF to avoid confounding the system, and to keep state [2] on until input I_3 occurs. This can be done through the use of active elements and a DELAY-IN component in the logic system as shown in Figure 8.2. Note that the circuit diagramed obviates the possibility of having more than one output state ON at any instant. Similar circuitry would apply to all the other states.

It is apparent from the logic diagram that, while the system can be implemented logically, it can also be awkward, inefficient, and potentially expensive. However, a theoretical basis for further analysis has been developed. Other methods of developing this logic will be shown in later sections.

8.2 COUNTING SYSTEMS

Binary coding, including binary counting codes, were superficially discussed in an earlier chapter.

One of the properties of an ordinal integer number, that is, an integer that shows the order or position of a set in a series of sets, is that each number represents a unique set. For example, the fifth set is a totally dif-

Sec. 8.2 Counting Systems

ferent set than the sixth! In fact, we choose to name the fifth set *five*, the sixth set *six*, and so on.

In the case of a decimal system, the sets named $n \times 10^m$, where n is an integer such that $0 \leq n \leq 9$ and m is any integer, are unique from each other. In the case of a binary system, those named $n \times 2^m$, where n is an integer such that $0 \leq n \leq 1$ and m is any integer, are unique. Under these conditions, if one can devise a scheme where it is possible to differentiate among various sets, one has created a condition where each combination of a unique *named* set AND the logic required for the desired operation of a system is in turn unique. Those devices that enable the use of this feature are counters, or electro-mechanical or pneumatic steppers. One literally counts events, or steps, through a process by noting the required logic for each state, and then taking the appropriate action. In effect, this is in reality an *event*-timed system, rather than a synchronous or clock-timed system.

8.2.1 Binary Counting Systems

Consider the element illustrated in Figure 8.3. Assume that when this device is turned ON, the *Y'* output is in the 1 state. If this is the case, then the *Y* output must be in the 0 state. The first time that the *T* (triggered) input is operated momentarily, or *pulsed*, the outputs change sense, that is, *Y'* becomes 0 and *Y* becomes 1. Note that these are not really simultaneous events, but as in the past, the time factor will be ignored since it is so small with respect to the system operation. Each subsequent pulse of the *T* input results in an output change. This particular device is called a *T Flip-Flop*

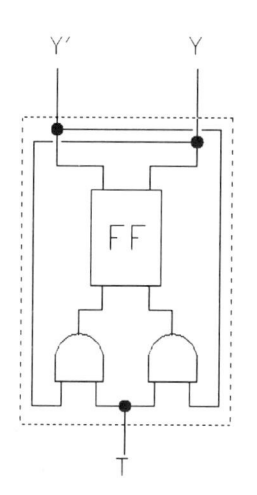

Figure 8.3a Schematic Representation. 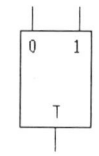   **Figure 8.3b** Symbol.

Figure 8.3 "T" Flip-Flop.

(TFF), and is constructed by combining a conventional, or *set-reset* (S-R) bistable with two AND gates as shown. This element is the core unit in a binary counter, and is illustrated in Figure 8.4 in conjunction with a binary-to-decimal decoder, resulting in a binary-to-decimal counter.

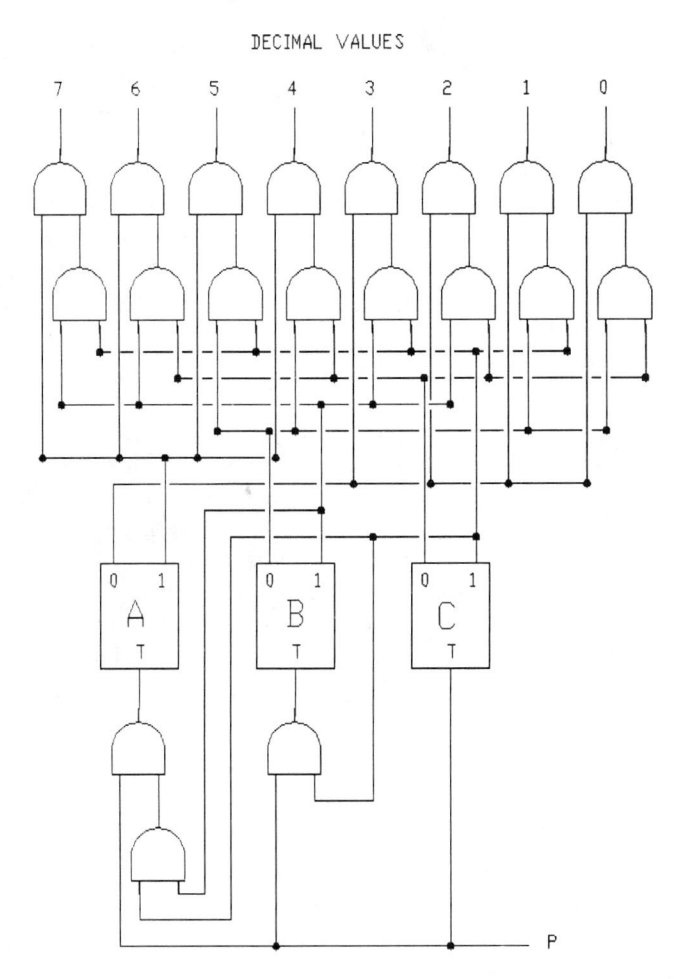

Figure 8.4 Binary to Decimal Counting System.

The system illustrated in Figure 8.4 shows an octal binary counter feeding a binary-to-digital converter or decoder, in which eight unique states can be differentiated and identified in decimal mode. It should be obvious to the reader that additional states could be accommodated simply by adding additional elements and interconnecting them appropriately.

First, consider the outputs of the three TFFs labeled A, B, and C in the binary counter illustrated at the bottom half of the diagram. As in the case of Figure 8.3, assume that all the binary outputs are initially in the 0 ON sense, and that P is a pulsed input. The output of the TFFs will be $ABC = 000$.

Upon the first input pulse at P, the TFF labeled C changes state, resulting in an output combination $ABC = 001$. In addition to this, a 1 signal appears at the inputs to the AND gates associated with TFFs B and C.

At the second pulse, the output of B changes state since the output of C is still ON, then C changes state so $ABC = 010$. Note that internal delays are required in order to insure the proper sequencing of these events. At the third pulse, B does not change state, but C does so $ABC = 011$.

In this manner, the output of the binary counter portion of this system will count from 0 through 7 (8 states). At the ninth pulse, the counter will reset itself to $ABC = 000$.

The upper section of this illustration shows the use of a series of AND gates to generate decimal code. Consider the situation when $ABC = 101$. Notice that the outputs of the three TFFs are connected in such a way that the third system output (from the left) is energized, while all the others are OFF. It should be apparent to the reader that $101_2 = 5_{10}$. Therefore, a decimal counter has been developed with unique sets for use with sequential systems. It should also be apparent that when this counter is in state $ABC = 111$ and a pulse appears at P, the system will revert back to state $ABC = 000$. An application of a counting system will be discussed later in this chapter.

8.3 ROTARY STEPPERS

This category of devices includes, but is not limited to electromechanical rotary stepping switches, rotary solenoids, and drum steppers, both electromechanical and pneumatic.

8.3.1 Rotary Stepping Switches

The rotary stepper switch (relay) was illustrated in Figure 5.3d and is repeated here in Figure 8.5. Since there are many different styles and configurations in this category, this discussion will be directed to the general group rather than a specific device.

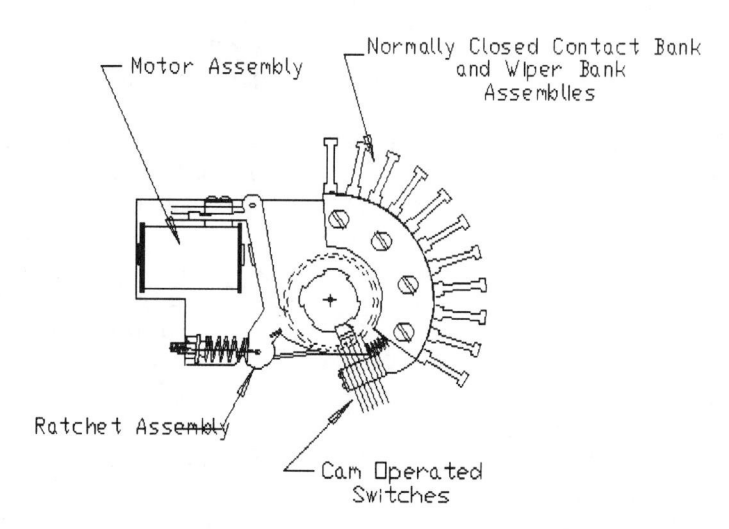

132 Sequential Logic Systems: Counting And Stepping Methods Chap. 8

Figure 8.5 Rotary Stepping Switch Assembly (Typical).

The initial application for the rotary stepping switch was as a rotary line switch. This device was used to electrically connect pairs of wires in telephone dial exchanges with other pairs in order to direct telephone calls from one party to another. Switching systems using these rotary steppers have been in service for well over fifty years with only routine maintenance required. It should be evident that these are extremely reliable devices and well suited to automation applications. Modern rotary steppers have evolved into devices that can utilize four different types of switching, which are illustrated in Figure 8.6.

Figure 8.6a shows some basic switching nomenclature that will be used in future discussions. In all the switching arrangements, the arrow above each element designates the direction of actuation of the motor blade. For ease in identifying these switching types, the following designations have been adopted.

Form *B*	*Break* element
Form *A*	*Make* element
Form *C*	*Make-then-Break* element
Form *D*	*Break-then-Make* element

Figures 8.6b and c show transfer wiper contact arrangements typical of those in use with rotary stepping switches. Figure 8.6b illustrates a *nonbridging* (NB) wiper where the brush spring ceases to touch one feeder contact prior to making contact with the next one, in effect a form *C* switching element. This arrangement is used when one wishes to transfer a circuit on a *discontinuity* basis. The other view shows a *bridging* (Be) arrangement

Sec. 8.3   Rotary Steppers

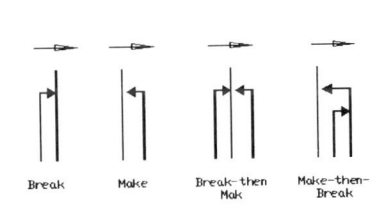

Figure 8.6a Switch Contact Form Nomenclature.

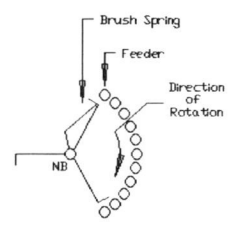

Figure 8.6b Non-Bridging Wiper Contacts (Break-then-Make).

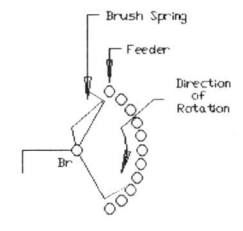

Figure 8.6c Bridging Wiper Contacts (Make-then-Break).

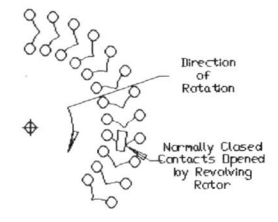

Figure 8.6d Portion of One Level of Normally Closed Contact Bank.

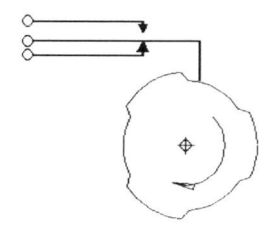

Figure 8.6e Cam Operated Switching.

Figure 8.6 Rotary Stepping Switch Details.

where the brush spring makes contact with the next contact prior to losing contact with the first, or a form D switching element. This arrangement is used when one wishes to transfer a circuit on a *continuity* basis.

In these two types of wiper contacts, the length of the end of the brush spring is the differentiating element. In the case of the Non-Bridging contacts, the end is significantly shorter, thereby indicating that only one feeder contact can be touched at any one time. Since this distinction is easily overlooked when viewing a diagram, the initials *NB* for Non-Bridging and *Be* for Bridging are commonly used, as is done in these drawings.

Figure 8.6d illustrates the third type of switching that can be done with rotary steppers. These *banks* consist of pairs of normally closed contacts that remain closed unless an insulated rotor is stepped into position in between a particular pair. In control applications, this switching can be used to move the rotor to a particular position, among other things.

Figure 8.6e shows the cam-operated contacts of a stepping switch. These operate in a binary mode; that is there are only two discrete positions, either *unoperated* or *operated*. These, along with the bridging and nonbridging contact banks, provide the primary switching that is used to activate the various devices in an automation system.

An illustrative example of the use of the general rotary stepping device will be presented later in this chapter.

8.3.2 Rotary Solenoids

A second type of electro-mechanical rotary stepper is the rotary solenoid, such as that supplied by Ledex Corporation of Dayton, Ohio.

These devices are true solenoids, or linear electro-mechanical devices, in that the basic movement of the plunger is linear, but due to the action of a cam on this linear motion, a slight rotation is imposed. This, coupled with a mechanical detent, results in a stepped rotation. If a rotary switch wafer is coupled with this solenoid, then an electrically stepped rotary switch is created which can be used as an automation controller. A typical rotary solenoid operated switch is illustrated in Figure 8.7. It should be noted that many wafer configurations, and steps per cycle, are available. Some of the most common are 10, 12, and 18 positions.

Sec. 8.3   Rotary Steppers   135

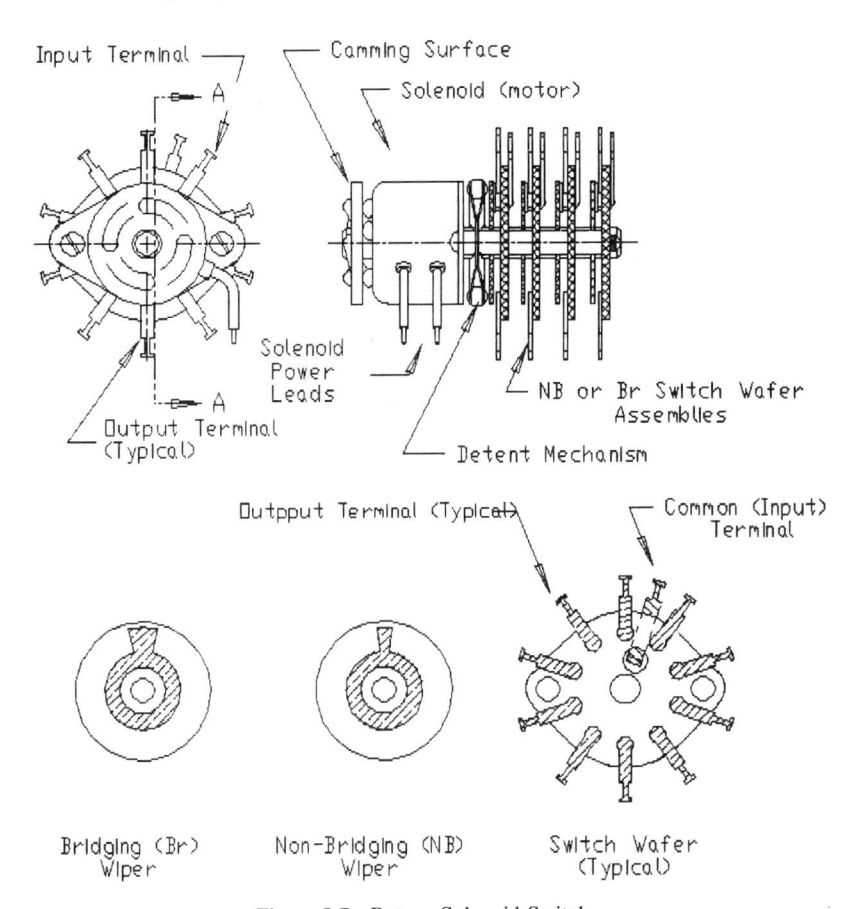

Figure 8.7 Rotary Solenoid Switch.

It should be noted that the switch wafers referred to in the above figure are similar to those illustrated in Figures 8.6b and c. Usually, one wafer acts as a solenoid controller, and the remaining wafers are assigned one per output.

8.3.3 Drum Steppers

This category of devices is very similar to the previously described devices in that there is a stepping motor that actuates switching elements, either electrical or pneumatic, to provide step programming. These devices are available from many sources. A typical drum stepper is shown in Figure 8.8.

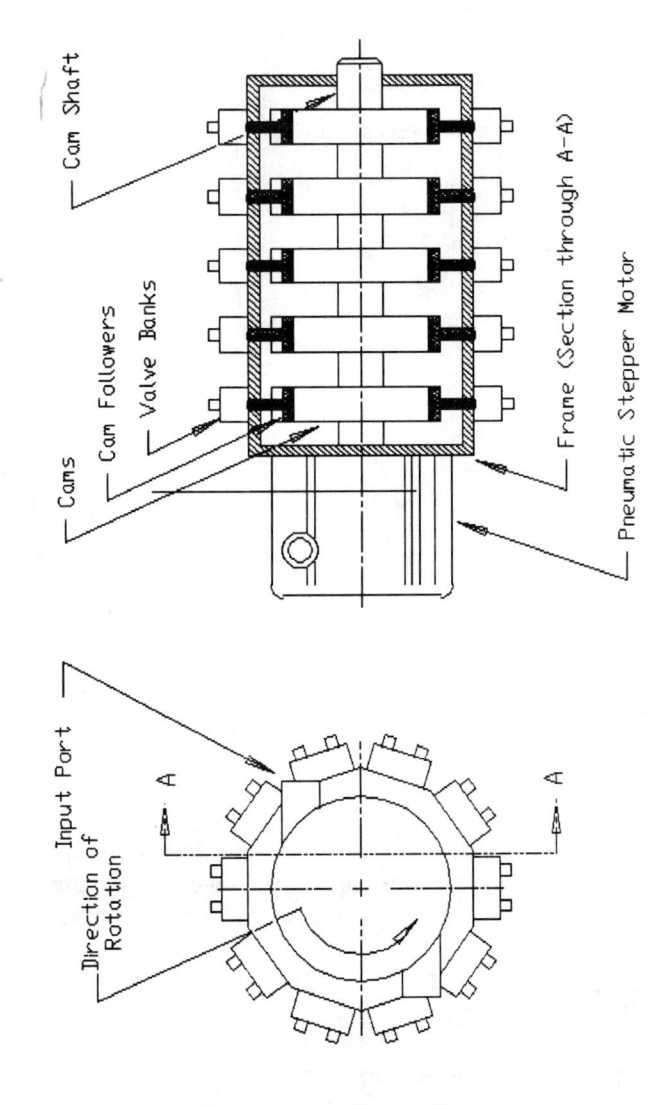

Figure 8.8 Pneumatically Driven Drum Stepper.

Sec. 8.4   Examples   137

The illustrated device shows a pneumatically driven stepper motor operating pneumatic valves. The principal of operation is essentially the same as that described in Section 8.3.2. These devices are usually, but not always, decimal devices with ten stepping positions. Single (as illustrated), or multiple lobed cams can be used. If a single lobed cam is used, a normally open valve is required at each step to generate an ON output signal for a single output variable. In other words, if, in a ten-step sequence, output Z_1 will be operated four times (in steps 1, 3, 6, and 7), four valves corresponding to this output will have to be provided. In a multiple lobed system, only one valve is required, but the cam would require lobes corresponding to each step that the output is to be ON.

8.4 EXAMPLES

Consider the system illustrated in Figure 8.1. This simple system could be automated with a control system employing a binary counter as shown in Figure 8.9, or a rotary stepper as illustrated in Figure 8.10. Since all the rotary stepping devices operate on exactly the same principal, only the rotary solenoid switch is shown. The use of a system employing a MEMORY has been demonstrated in Chapter 7, Figure 7.5.

Consider the shaded area in Figure 8.9. It can readily be seen that this

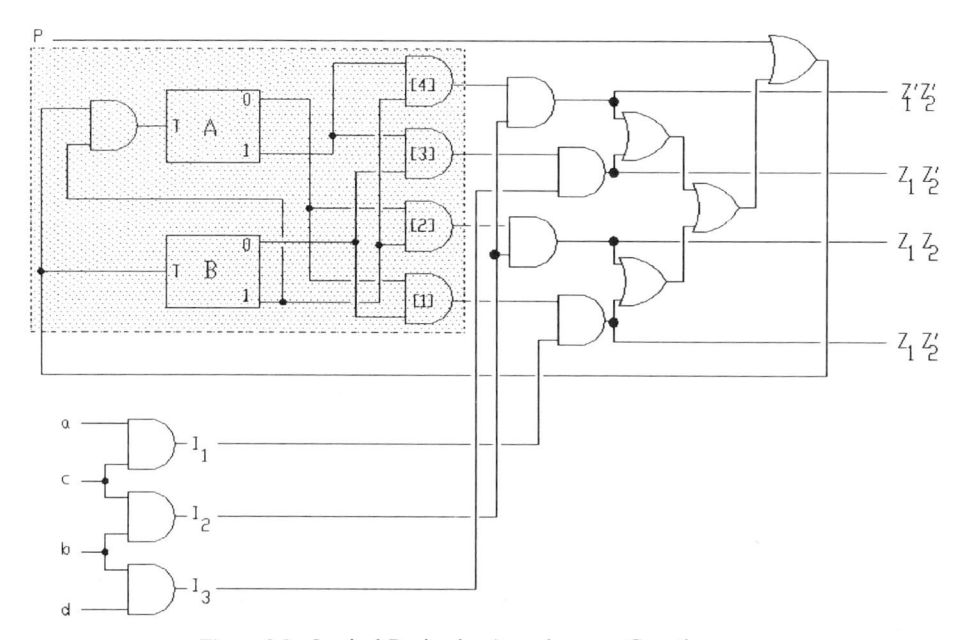

Figure 8.9 Logical Design by Asynchronous Counting.

is nothing more than a four-level decimal-coded binary counter. Below it are three AND gates which provide the output coding for the three input statements I_1, I_2, and I_3. As mentioned before, the automation design technique is to couple the appropriate input combination to the correct counter state, thereby generating a unique overall input condition.

A counter system must always have two sets of outputs. The first, of course, consists of the required system outputs. The second output set provides the signal to *step* or advance the counter to the next state. For example, if the system is in state $AB = 00$ and inputs a and c both turn ON, then the output will become $Z_1Z_2 = 10$, and a signal will be generated at P to step the counter. The reader should verify that the other required output conditions are provided. Attention should be paid to the fact that the number of OR gates that are used to enable the *step* signal is necessitated by the presence of five inputs to two-input devices.

This system illustrates all the required sub-systems of a counter controlled automation scheme:

- a decimal coded binary counter,
- a logical input sub-system that combines the system inputs in an appropriate manner, and finally
- a logical sub-system that combines the other two to provide the appropriate output signals.

Automation using the properties of the various rotary stepping devices is basically identical. It is for this reason that only the Rotary Stepping Switch is illustrated.

As in a counter controlled automation system, a stepper controlled system has three essential parts:

- a piece of hardware, or a hardware system that will provide unique and regular *steps* (states),
- a way of controlling the stepping of the hardware system, and
- a method of generating the appropriate system outputs.

The reader's attention is directed to Figure 8.10. A rotary solenoid switch with three wafers is illustrated, as are schematic diagrams of the various interconnections required. The three wafers are labeled 1, 2, and 3, and are all of the Non-Bridging variety. Wafer 1 acts as the controller for the stepping of the solenoid, wafer 2 generates the output signals for output Z_1, and wafer 3 provides the output Z_2.

Consider that the three wafers are interconnected in such a way that all are in the same position at any given time. That is, if wafer 1 is in position 3, then wafers 2 and 3 are also in position 3, and so on. Position 1 is the

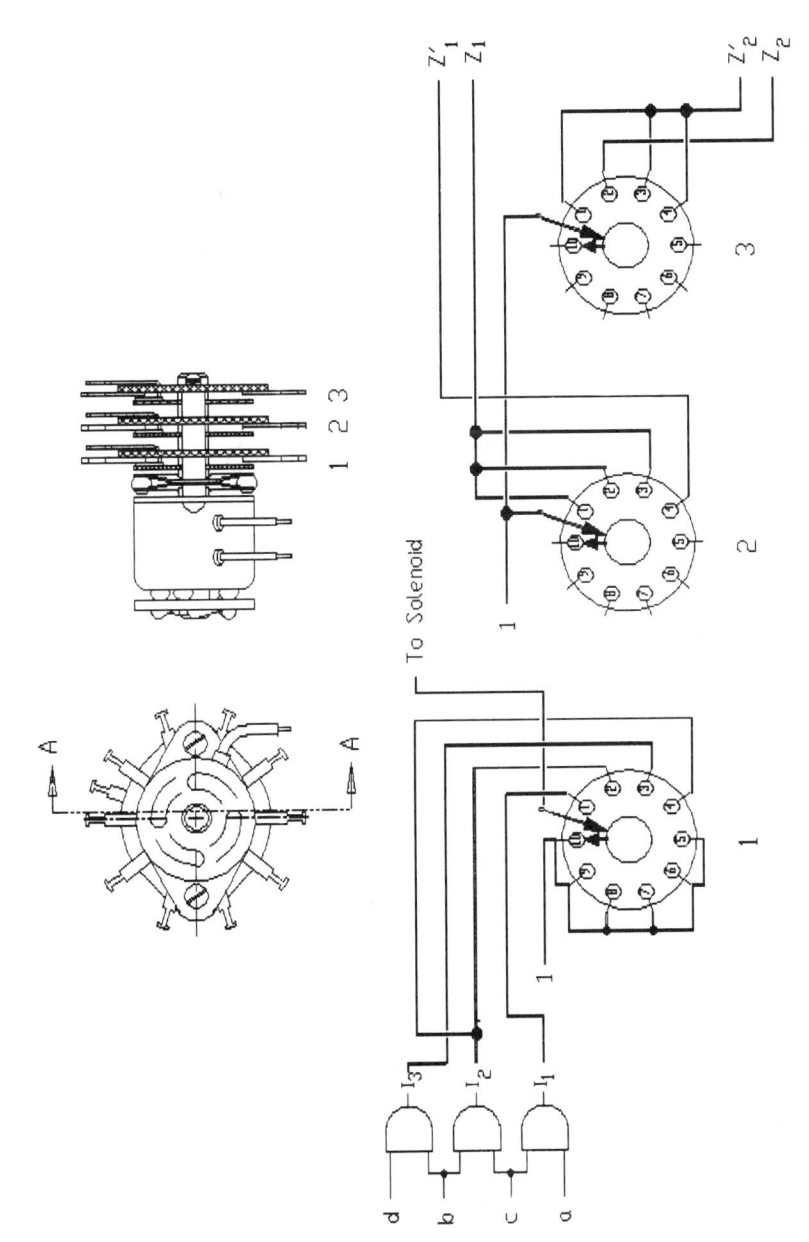

Figure 8.10 Logical Design Using Rotary Stepping Switch.

normal starting position. Think about what is happening at each step of the driving solenoid.

If the device is in position 1, and $ac = (11)'$, then the output signal from wafer 2 is Z_1, and from wafer 3, Z_2'. As soon as the input changes to $ac = 11$, then a signal passes to the solenoid causing it to step to the next position, and changing the outputs to $Z_1Z_2 = 11$.

In a manner similar to the sequence of events described above, if the control device is now in position 2, and the inputs $bc = 11$, then the solenoid is again energized, the device steps to the next position (position 3), and the outputs $Z_1Z_2 = 01$.

This sequence of events continues to position 4. When the inputs are $bc = 11$, the stepper goes to position 5, but position 5 is so configured so that a signal is immediately sent to the solenoid, which steps the device to position 6. Eventually, the stepper reaches position 1 again, and stays there (or *homes* in at that station) until the appropriate input combination exists.

A very cursory examination of these illustrations shows that in this case, that is one with relatively few input combinations and stable states, the use of a MEMORY will *probably* result in the optimal system. However, any of the other schemes could also be effectively used. Other considerations, such as system complexity, response time, reliability, maintainability, and required service life, will certainly affect any ultimate design decision.

As stated before, this example uses a very simple system where there is really no clear-cut advantage to the use of a MEMORY. Consider, however, the system illustrated in Figure 7.9, which differs from the simple sequencing system *only* through the addition of a power failure safety feature.

The number of input combinations is now eleven rather than three, and the number of stable states is twelve rather than four. Since there are eleven inputs and twelve stable states, there must be some stable states that are confounded. On examination of the LSC, it can be seen that the two states in question are [3] and [5]. As shown in Figure 7.5, these two states can be differentiated by the use of a single MEMORY.

Compare this with a logical decimal coded binary system, where a hexadecimal digital coded binary counter would have to be provided in addition to a series of AND gates. The AND gates can be simplified from five-input to three-input by using the fact that both A and B cannot be ON at the same time, nor can C and D. This logic would, of course, be necessary in any automation scheme (unless some reductions were possible). But it should be remembered that one of the features of a counter/stepper automation system is the fact that the system can be developed from the original LSC without any further logical reduction.

Instead of using either the single MEMORY or the decimal coded binary counter, consider the use of a stepping device. There are only two output states; so, in the case of the rotary solenoid switch, only a solenoid controller and two output wafers would be required, but a unit that steps *at least* eleven times would be needed. Assuming that either a twelve or eighteen step device were available, automation could be accomplished through the use of the logical inputs to the solenoid coupled with a single rotary solenoid switch.

In this case, there appears to be an advantage to the use of the single MEMORY device automation system.

In the case of a system such as is illustrated in Figure 7.10, in order to implement this automation system with no further logical reduction, eight MEMORY elements, a hexadecimal decimal coded binary counter, or a single eighteen step rotary solenoid with three wafers, would be required. Here, chances are that the rotary stepper would be the hardware system of choice.

To recapitulate, the biggest advantage of counting and stepping systems lies in the ease of design resulting from their use. There is no need to combine states and reduce the logic from that stated in the original Logic Specification Chart. While this may result in a system that is inoptimal from the hardware standpoint, the simplicity of the engineering and design coupled with the preservation of the elementary design logic can result in a system that is both effective and mnemonic. This reduces design cost as well as providing a controller that is easily serviced.

8.5 PROBLEMS

1. Design a hexadecimal output digital coded binary counter using logical elements only.
2. Design a digital coded binary counter automation system for the press safety system of problem 6.2. Use the PFT generated in problem 7.1. Do you feel that this is an appropriate choice for an automation technique? Defend your position.
3. Design a rotary stepper system for the safety device described in chapter 6, problem 6.2. Use the PFT generated in Problem 7.1. Do you feel that this is an appropriate choice for an automation technique? Defend your position.
4. Using the LSC developed in problem 7.4, design an automation system using counting/stepping techniques.
5. Design a binary counter automation system to implement the LSC shown in Figure 7.9.
6. Design a binary counter automation system to implement the LSC shown in Figure 7.10.
7. Design a binary counter automation system to implement the RSC shown in Figure 7.15.
8. Design a stepper automation system to implement the LSC of problem 7.10.
9. Design a stepper automation system to implement the LSC of problem 7.11.
10. Design a stepper automation system to implement the LSC of problem 7.12.

Chapter 9

SEQUENTIAL LOGIC SYSTEMS: PASSIVE MEMORY TECHNIQUES

As opposed to the techniques discussed previously, the passive memory methods described in this chapter are time-tested, reliable schemes. They have been used in industrial automation systems for many years. They have many advantages and are readily implemented.

A passive memory is one that requires no energy to maintain either of its two MEMORY states. This means that when switching from one state to the other, energy is required; but as long as there is to be no change in state, the devices *remember* and maintain their last position.

In order to accomplish this, MEMORY devices rely on mechanisms of some sort to retain their current state. These elements were described in Chapter 5. They include mechanical, electro-mechanical, and MPL devices.

9.1 PASSIVE MEMORY HARDWARE

Any memory device which does not rely on an *active* power source to retain its current output state is said to be a *passive* MEMORY element. In most cases, these components have a mechanical memory. The logic signal is directed through the device according to the mechanical position. An example of this is the four-way two-position valve shown in Figures 5.6 and 5.9.

The most important feature of these devices is their mechanical memory. Once the valve has been switched by either the SET or RESET signals, it remains in that position due either to friction or to a detent. The signal sent through the device does not necessarily have to be an active signal connected to the supply; it may be an input or a logic signal which appears only occasionally.

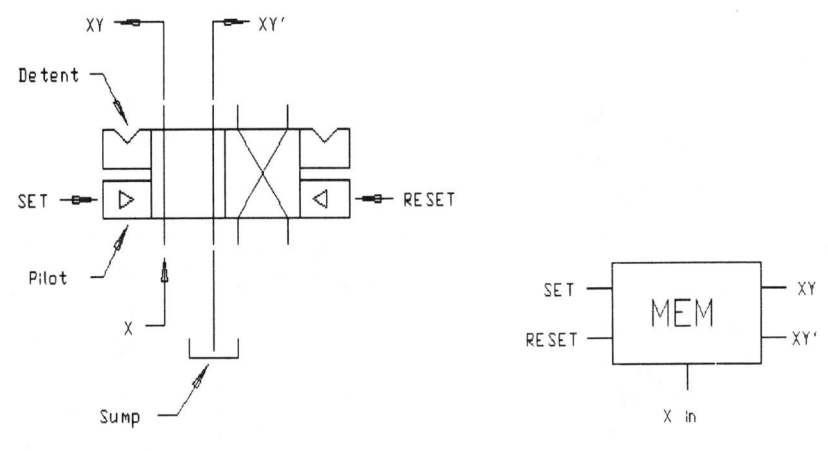

Figure 9.1a Standard Symbol.   **Figure 9.1b** Simplified Symbol.

Figure 9.1 Passive MEMORY Valve.

Sec. 9.1 Passive Memory Hardware

When a logic signal acts on a passive MEMORY device, the output XY appears only when the memory element is in the proper position (indicated in the illustration by Y) AND the logic signal X is ON. This device holds its mechanical position while it forms one of two AND combinations, XY or XY'. The passive memory device therefore serves the function of three logic elements, a MEMORY and two ANDs. By using this feature, logical automation systems can be simplified and hardware requirements can be significantly reduced.

Another feature of this device is the complementary output. Note that the device has two outputs XY and XY'; when one is ON, the other is OFF. This eliminates the need to invert Y in order to obtain its complement Y'.

Fluidic passive MEMORY devices, without moving mechanical parts, do not exist.

Electro-mechanical devices are available in a configuration that provides a passive MEMORY. The mechanical latching relay provides a passive MEMORY. Another passive MEMORY device is the double-pole double-throw over-center double-wound reed relay illustrated in Figure 9.2. Note that the direction of winding of the *armature coils* are opposite to each other, so an electromotive force is created in one winding which results in the movement of the armature blade in one direction, where it stays due to its

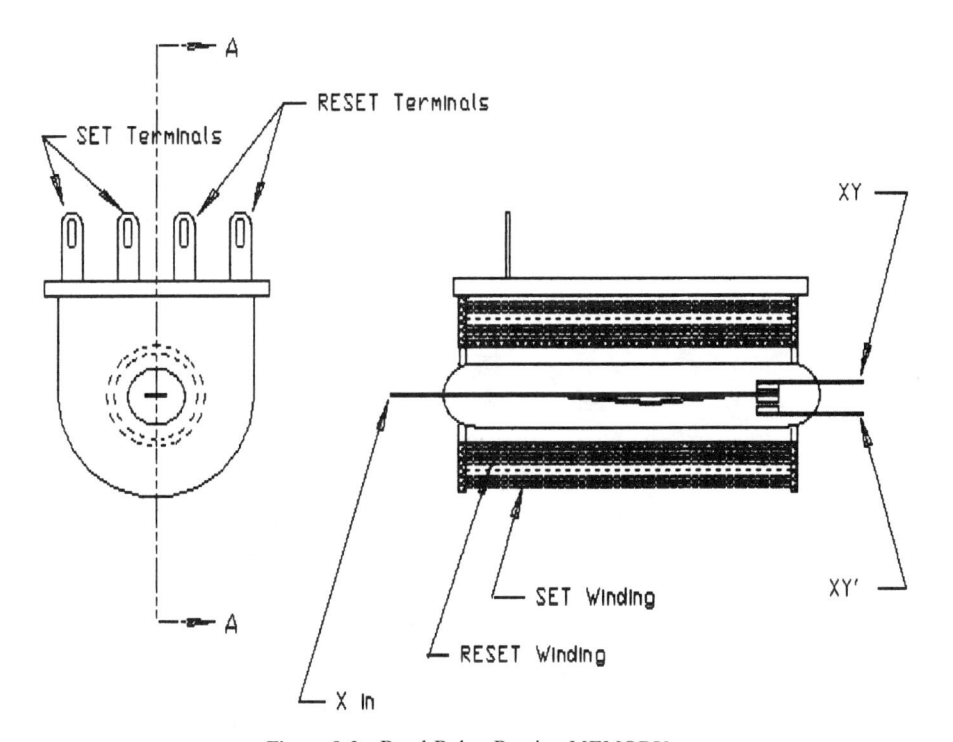

Figure 9.2 Reed Relay Passive MEMORY.

over-center geometry until the other coil is energized. This in turn causes the armature blade to move to the other pole. Again, the output of this device is XY or XY' as with the four-way valve.

The passive memory can be used to reduce the amount of required hardware when distinguishing between two states. The problem of assignment when higher orders of memory are required will be discussed later in this section. For example, by using two passive MEMORY elements, three unique MEMORY states can be developed, as shown in Figure 9.3a. Four unique states are obtained by using three MEMORY elements as shown in Figure 9.3b., and so on.

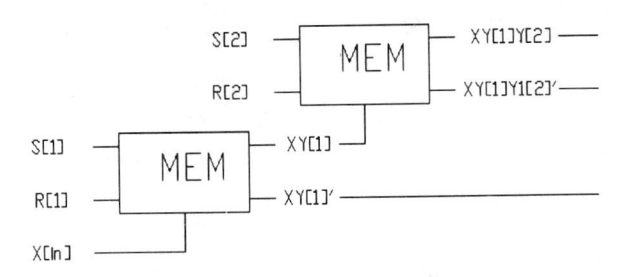

Figure 9.3a 3 Unique Output States.

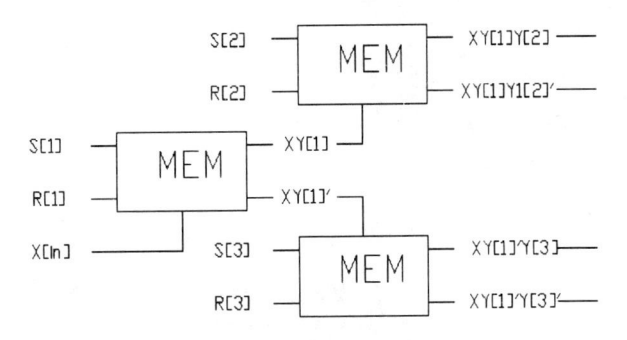

Figure 9.3b 4 Unique Output States.

Figure 9.3 Passive MEMORY Assignment Schemes.

Each time a MEMORY element is added to the system, another unique passive memory state is created. In general, $n - 1$ passive MEMORY elements describe n unique MEMORY states. The assignment plan in Figure 9.4 shows a passive memory code. As stated above, to describe n unique states, omit all MEMORY elements numbered above $n - 1$. The alternating placement of elements in this assignment code allows the proper balancing of power flow. Still higher order assignments may be obtained using a similar alternating pattern.

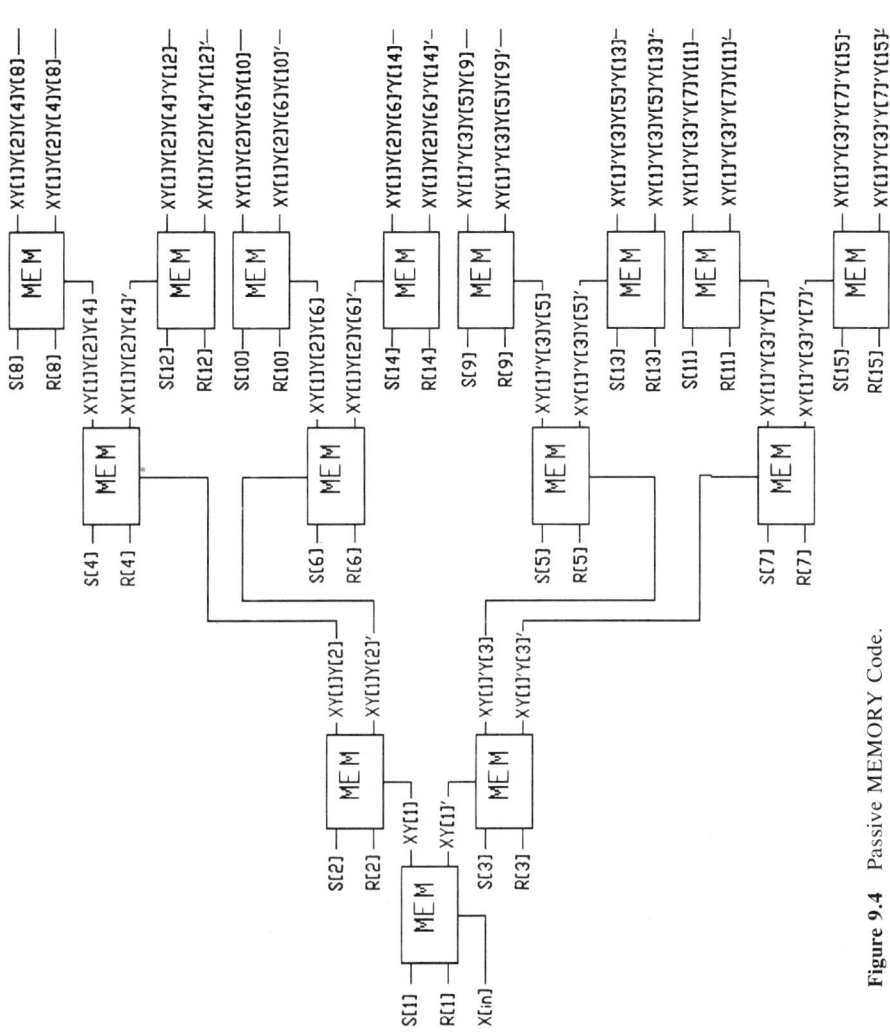

Figure 9.4 Passive MEMORY Code.

9.2 STOCHASTIC SYSTEMS

The assignment procedure for a stochastic system can be demonstrated by another example. Consider the PFT for the system shown in Figure 9.5.

INPUT STATES	In	NEXT STATES			OUTPUT STATES
n	1	2	3	4	
a	0	1	1	0	Z
b	0	0	1	1	
	[1]	2	-	5	0
	1	[2]	3	-	0
	-	7	[3]	4	0
	1	-	6	[4]	0
	1	-	6	[5]	1
	-	7	[6]	5	1
	1	[7]	6	5	1

Figure 9.5 Primitive Flow Table Stochastic System.

The synthesis of a system to perform the logic represented by Figure 9.5 proceeds directly from the PFT, (or LSC or RSC as applicable). Notice that stable state [1] is a source state, that is a state that is the *only* stable state in a given column of a PFT, LSC, or RSC. For this reason, state [1] can always be uniquely represented by the input combination $A'B'$. This is not the case for the input combination of the second column. Either of two events could possibly occur when the input combination AB' appears. One time, the output is OFF (state [2]), while the other time the output is to be ON (state [7]). Thus, some method to distinguish, or discriminate, between the two states must be employed. This is most conveniently handled through the use of a passive MEMORY element. Having the MEMORY in the SET condition for one state and in the RESET condition for the other allows each of the states to be unique. This signal is the combination of the input state and the memory state. Specifically, the assignment of the RESET memory state, Y_{1R}, to state [2] and the SET state, Y_{1S}, to state [7] results in unique state signals for both states. The input signal for state [2] is now $AB'Y_{1R}$, and that for state [7] is $AB'Y_{1S}$.

Observation of the third column reveals a similar problem. States [3] and [6] also have to be made unique through the assignment of memory states. To accomplish this, RESET memory Y_{2R} is assigned to state [3], and Y_{2S} to state [6]. State [3] is now represented uniquely by ABY_{2R} and state [6] by ABY_{2S}.

The input signals for states [4] and [5] in the fourth column also require

Sec. 9.2 Stochastic Systems   149

uniqueness, which can be accomplished through the use of a third MEMORY, Y_3. This permits state [4] to be represented by $A'BY_{3R}$, and state [5] by $A'BY_{3S}$.

In order to insure that the MEMORY elements are in the proper state, the switching conditions for SET and RESET must be specified. For example, Y_1 must be in the RESET state for state [2]. Therefore, Y_1 can be reset by any state prior to state [2]. In this case, state [1] is the only one providing a transition path to state [2], so Y_1 is RESET by the signal associated with state [1]. Since Y_1 must be in the SET state for state [7], it can be switched by any state prior to state [7]. The transition paths shown in Figure 9.5 show that either state [3] or state [6] has a possible transition path to state [7], therefore either could be used for the required SET.

Similar reasoning can be used to develop the switching conditions for MEMORIES Y_2 and Y_3. The switching conditions are summarized in Figure 9.6. Notice that it is often possible to reduce and simplify the expressions.

	RESET	SET
Y[1]	State [1] = A'B'	States [3] or [6] = ABY[2R] + ABY[2S]
Y[2]	State [2] = AB'Y[1R]	States [4], [5], or [7] = A'BY[3R]+A'BY[3S]+AB'Y[1S]
Y[3]	State [3] = ABY[2R]	States [1] or [6] = A'B'+ABY[2S]

Figure 9.6 Memory Switching Conditions for System of Figure 9.5.

The switching of MEMORY elements prior to their required state provides the prepared flow path feature, in that the memory state for either transition from a stable state is already prepared. This eliminates the delay introduced by switching the MEMORY after an input change is encountered.

Notice that since the MEMORY elements prepare their own paths, it is only necessary to initially set the MEMORY states for the starting of the system's operation. In this case, state [1] requires no memory, and states [2] and [5] are prepared by state [1]; therefore no MEMORY initialization is required.

At this point, each state in the PFT has a unique signal representation, as was the case with the counting/stepping systems described in Chapter 8. Proper care must be taken to ensure that the MEMORY elements will be in the right states at the appropriate times. The output signals can now be generated by combining the existing state signals. By referring to Figure 9.4 it can be readily seen that the output Z appears at states [5], [6], or [7]. The output equation can therefore be generated by combining these state signals as:

$$Z = [5] + [6] + [7] = A'BY_{3S} + ABY_{2S} + AB'Y_{1S}$$

Specification of the state signals, switching conditions, and the output equations as given below completes the design procedure. Notice that it is possible to use only the state numbers rather than the actual state signals when specifying the switching and output equations.

State Signals: $[1] = A'B'$
$[2] = AB'Y_{1R}$
$[7] = AB'Y_{1S}$
$[3] = ABY_{2R}$
$[6] = ABY_{2S}$
$[4] = A'BY_{3R}$
$[5] = A'BY_{3S}$

Memory States: $Y_{1R} = A'B'$
$Y_{1S} = AB$
$Y_{2R} = AB'Y_{1R}$
$Y_{2S} = A'B + AB'Y_{1S}$
$Y_{3R} = ABY_{2R}$
$Y_{3S} = A'B' + ABY_{2S}$

Output Equation: $Z = [5] + [6] + [7]$

The schematic logic diagram is shown in Figure 9.7.

Sec. 9.3   Deterministic Systems   151

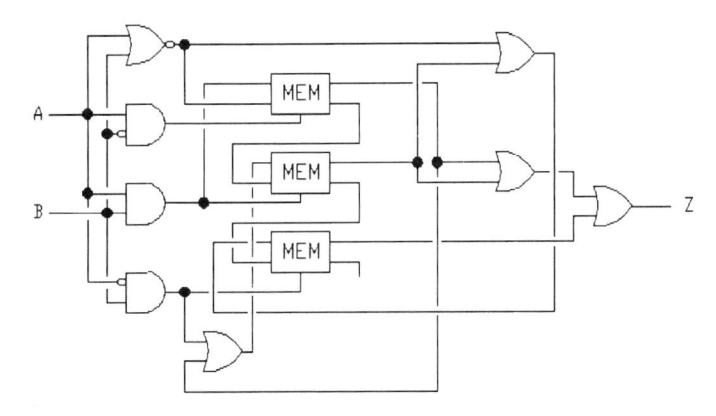

Figure 9.7 Schematic Logic Diagram for System of Figure 9.5.

One further consideration should be noted. The system that has just been developed may not be *truly* optimal. For example, it is possible to develop four unique memory states by using only two MEMORY elements concurrently, for example, $Y_{1R}Y_{2R}$, $Y_{1R}Y_{2S}$, $Y_{1S}Y_{2S}$, and $Y_{1S}Y_{2R}$. This procedure has some significant drawbacks in that additional logical ANDs are required, and appropriate SET and RESET signals can be extremely difficult to generate.

9.3 DETERMINISTIC SYSTEMS

As was discussed previously, deterministic systems are characterized by a regular, repetitive action resulting in one and only one prior state for each stable state in the system.

Consider the deterministic system shown in the LSC in Figure 9.8. This is representative of a two pneumatic cylinder extend/retract deterministic sequence: E-1, E-2, R-2, E-2, R-1, and R-2, where E represents *extend* and R represents *retract*.

	NEXT STATES					OUTPUT STATES
	-------------	---	---	---	---	---------------
	INPUT STATES: In					
EVENT	n	1	2	3	4	$Z_1 Z_2$
	a	0	1	1	0	
	b	0	0	1	1	
E-1		[1]	2	-	-	10
E-2		-	[2]	3	-	11
R-2		-	4	[3]	-	10
E-2		-	[4]	5	-	11
R-1		-	-	[5]	6	01
R-2		1	-	-	[6]	00

Figure 9.8 Primitive Flow Table Deterministic System.

The synthesis of the system needed to perform the logic represented by Figure 9.8 can also be accomplished by direct analysis of the LSC. Note that stable states [1] and [6] are source states, and can be uniquely represented through the use of the input combinations $A'B'$ and $A'B$ respectively. On inspection of these combinations, it appears that this logic could be implemented through the use of NOR and INHIBIT elements; or by inputting the complement of A, and using AND and INHIBIT devices. However, since the balance of the system must be analyzed and synthesized prior to making this decision, it would be premature to select hardware at this time.

The second column of the PFT represents an entirely different situation (when the input combination is AB'). This input combination leads to either stable state [2] or [4], which appears to be confounded. Upon further analysis, it is noted that the outputs for the two stable states are identical, that is Z_1Z_2, which leads to the consideration of combining these two states into a new state. Then it is noted that the next state for [2] is [3] whose output is Z_1Z_2', while the next state for [4] is [5] where the output is $Z_1'Z_2$. This is, of course, an analysis similar to that shown in Chapter 6 in order to develop a RSC.

Since these two states cannot be merged into one, one can discriminate between them through the use of a passive MEMORY. As with a stochastic system, this can be accomplished by having the MEMORY element SET for one state, and RESET for the other.

Analysis of the third column of the PFT results in a similar condition with stable states [3] and [5]. Here the outputs are different, so there can be no further question as to whether the two states can be merged.

Following the procedure that was demonstrated for use with stochastic systems, two passive MEMORY elements will be used. The first will dis-

Sec. 9.3   Deterministic Systems 153

criminate between stable states [2] and [4], the second between stable states [3] and [5]. The first MEMORY will be RESET at state [1] (since it is the state prior to state [2]), and SET at state [3]. The second MEMORY *could* be RESET at state [2] (following the above procedure) or at any state between the start of the sequence and state [2] (in this case, state [1]), and SET at state [4]. A problem exists, however, if MEMORY 2 is RESET at state [2] ($AB = 10$) and SET at state [4] ($AB = 10$). Obviously, the same input cannot be used to both SET and RESET a MEMORY! Therefore the RESET will have to be performed at [1] ($AB = 00$) and the SET at [4] ($AB = 10$). The system equations are listed below.

State Signals: $[1] = A'B'$

$[2] = AB'Y_{1R}$

$[3] = ABY_{2R}$

$[4] = AB'Y_{1S}$

$[5] = ABY_{2S}$

$[6] = A'B$

Memory States: $Y_{1R} = A'B'$

$Y_{1S} = AB$

$Y_{2R} = A'B'$

$Y_{2S} = AB'$

Output Equations: $Z_1 = [1] + [2] + [3] + [4]$

$= A'B' + AB'Y_{1R} + ABY_{2R} + AB'Y_{1S}$

$= A'B' + AB' + ABY_{2R} = B' + AY_{2R}$

$Z_2 = [2] + [4] + [5] = AB'Y_{1R} + AB'Y_{1S} + ABY_{2S}$

$= AB' + ABY_{2S} = A(B' + Y_{2S})$

Notice should be paid to the fact that even though two passive MEMORIES are required to discriminate among *all* the states, the combination of next state conditions and output assignments are such that the first MEMORY is unassigned and, therefore, is not needed for the implementation of the outputs of this system.

Once the system equations have been written, hardware can reasonably be selected. Considering the equations required to describe this system, it can be readily seen that by inputting the complement of B, only five logic elements, including one Passive MEMORY component, are required for implementation. This system is illustrated in Figure 9.9.

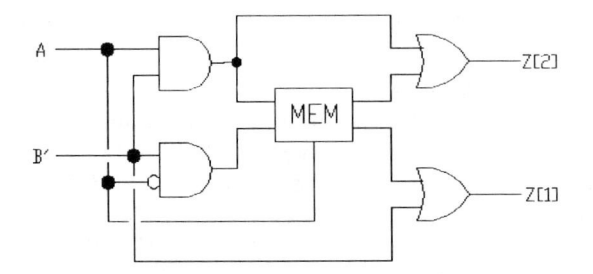

Figure 9.9 Schematic Diagram for $Z[1] = B' + AY[R]$; $Z[2] = AB' + AY[S]$.

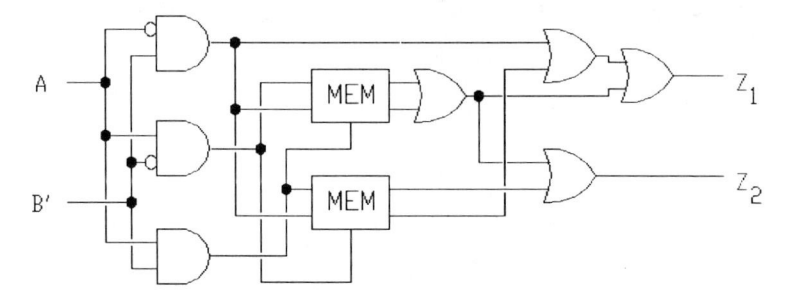

Figure 9.10 Hazard Free Implementation $Z1 = A'B' + AB'(Y[1R] + Y[1S]) + ABY[2R]$; $Z2 = AB'(Y[1R] + Y[1S]) + ABY[2S]$.

9.4 PROBLEMS

1. Write the system equations and develop a schematic diagram for the solution of the automation problem defined in the PFT of Figure P9.11.
2. Write the system equations and develop a schematic diagram for the solution of the automation problem defined in the PFT of Figure P9.12.
3. Write the system equations and develop a schematic diagram for the solution of the automation problem defined in the PFT of Figure P9.13.
4. Write the system equations and develop a schematic diagram for the solution of the automation problem defined in the PFT of Figure P9.14.
5. Write the system equations and develop a schematic diagram for the solution of the automation problem defined in the PFT of Figure P9.15.
6. Write the system equations and develop a schematic diagram for the solution of the automation problem defined in the PFT of Figure P9.16.

Sec. 9.4 Problems

INPUT STATES: In		NEXT STATES			OUTPUT STATES
n	1	2	3	4	Z_1Z_2
a	0	1	1	0	
b	0	0	1	1	
[1]	2	6	3		00
5	[2]	10	–		01
1	–	10	[3]		-1
1	[4]	9	3		01
[5]	4	10	7		00
–	4	[6]	3		11
5	–	6	[7]		01
1	–	9	[8]		10
–	2	[9]	3		11
–	2	[10]	7		1-

Figure P9.11

INPUT STATES: In		NEXT STATES			OUTPUT STATES
n	1	2	3	4	Z_1Z_2
a	0	1	1	0	
b	0	0	1	1	
[1]	2	5	–		00
1	[2]	–	8		01
7	–	[3]	9		10
[4]	6	3	–		11
1	–	[5]	9		10
4	[6]	–	8		01
[7]	2	3	–		00
–	6	3	[8]		11

Figure P9.12

INPUT STATES: In		NEXT STATES			OUTPUT STATES
n	1	2	3	4	Z_1Z_2
a	0	1	1	0	
b	0	0	1	1	
[1]	2	–	4		01
1	[2]	3	–		00
–	2	[3]	4		00
7	–	5	[4]		11
–	6	[5]	–		01
1	[6]	–	–		10
[7]	8	–	9		01
7	[8]	–	9		11
–	2	–	[9]		01

Figure P9.13

INPUT STATES: In		NEXT STATES			OUTPUT STATES
n	1	2	3	4	Z_1Z_2
a	0	1	1	0	
b	0	0	1	1	
[1]	3	–	2		10
4	–	5	[2]		10
1	[3]	–	–		10
[4]	3	–	2		10
–	–	[5]	6		01
1	–	–	[6]		01

Figure P9.14

	NEXT STATES								OUTPUT STATES
	INPUT STATES: In								
n	1	2	3	4	5	6	7	8	$Z_1 Z_2 Z_3$
a	0	1	1	0	0	1	1	0	
b	0	0	1	1	1	1	0	0	
c	0	0	0	0	1	1	1	1	
[1]	2								000
1	[2]	3							010
		[3]	4						010
			5	[4]	6				110
			7	[5]					110
					[6]	8			001
1	[7]					9	[8]	10	100
						[9]	8		001
							11	[10]	100
						12	[11]		001
						[12]			110
								13	110
1	14							[13]	010
	[14]								010

Figure P9.16

	NEXT STATES								OUTPUT STATES
	INPUT STATES: In								
n	1	2	3	4	5	6	7	8	$Z_1 Z_2 Z_3$
a	0	1	1	0	0	1	1	0	
b	0	0	1	1	1	1	0	0	
c	0	0	0	0	1	1	1	1	
[1]	2	–	3	–	–	–	–	4	000
1	[2]	5	–	6	–	–	–	–	001
1	–	7	[3]	8	–	–	–	–	011
1	–	–	–	9	–	–	10	[4]	010
–	–	11	[5]	12	–	–	–	13	100
–	–	3	[6]	14	–	–	4	–	101
1	[7]	3	–	–	–	–	–	–	110
–	–	–	12	[8]	14	–	–	13	111
–	–	–	3	[9]	14	–	–	4	111
–	–	–	–	14	[10]	13	–	–	011
1	[11]	7	–	–	–	–	10	–	001
–	2	–	–	–	–	–	–	–	100
1	–	7	[12]	8	–	–	–	–	100
1	–	–	–	–	–	–	10	[13]	010
–	–	5	–	6	[14]	10	–	–	001
–	–	–	–	–	–	–	–	–	000

Figure P9.15

Chapter 10

SEQUENTIAL LOGIC SYSTEMS: ACTIVE MEMORY TECHNIQUE

The counting/stepping and passive MEMORY automation techniques that were presented in the preceding two chapters are workhorses in that they have been used extensively and successfully in the past. The active memory technique that will be discussed here is also a viable scheme. It is one of great power resulting in logically optimal automation systems. It is widely used in the design of automation systems. However, it should be kept in mind that there is no single synthesis that will *always* prove superior or provide a universally most desirable system. As with many other disciplines, the selection of a method, or approach, is at best a compromise.

It is appropriate to restate the overlying principle of engineering design at this time:

> The optimum engineering design is one that results in a system that performs its design mission exactly, no more and no less, as reliably as necessary, at the lowest *overall* cost.

All the schemes presented here will, if properly executed, result in an automation procedure that will meet its functional goal. However, in order to provide a truly optimal system, one must balance factors concerning ease of design, reliability, maintainability, and implementation.

The major advantage of counting and passive MEMORY schemes is the ease and simplicity with which they can be applied. They probably will *not* result in logically optimal systems, but the logic that was used in their implementation is not lost in the design process. In the case of the classical active MEMORY scheme, an optimal system can be generated. However, due to the many minimization steps that are used in the process, the basic logic becomes obfuscated. This results in a system that usually cannot be serviced by logically tracing the function through to its conclusion. In this case, one must not only design the operating system, but must also specify and design a related maintenance program. Remember that all systems will fail in time; the question is *when*!

10.1 ACTIVE MEMORY HARDWARE

The MEMORY units described in the previous chapter, Section 9.1, represent hardware that can be used in either active or passive MEMORY applications. Passive components such as four-way double-piloted valves and maintaining-contact relays can be made active merely by applying a source to the X input. Other active elements, that is those that inherently require a power source due to the presence of a complementing input or output (such as fluidic and electronic bistable devices) are regularly employed.

The primary difference between active and passive components is one

Sec. 10.1   Active Memory Hardware 159

of how the device is installed. In the case of an active MEMORY, a separate power source is needed in order to maintain the output. In the case of a passive MEMORY, the output power derives from the ANDing of an input with the MEMORY. This difference is schematically illustrated in Figure 10.1 using as examples the Flip/Flop and MEMORY elements. Note the presence of a source of flow is *always* understood to be present in a Flip-Flop.

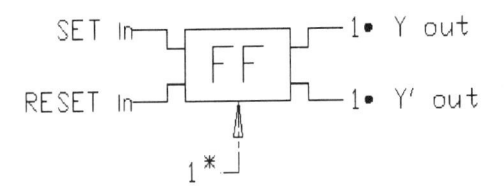

```
SET In ──┐         ┌── 1• Y out
         │   FF    │
RESET In ┘    ↑    └── 1• Y' out
             1*
```

* The source "1" is always understood to be present in a bistable element (but never drawn)

Figure 10.1a Bistable (Flip Flop) "Active" Element.

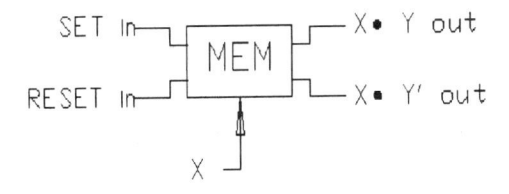

```
SET In ──┐         ┌── X• Y out
         │   MEM   │
RESET In ┘    ↑    └── X• Y' out
              X
```

Figure 10.1b MEMORY "Passive" Element.

Figure 10.1 FF/MEM Comparison.

In the case of active components, logical devices can be constructed using NOR or INHIBIT elements. This is shown schematically in Figure 10.2. To implement this technique, specific devices can be configured to provide the bistable action, such as the two three-way single-piloted valves shown in Figure 5.9, the NOR fluidic element illustrated in Figure 5.8e, and the multiple diaphragm device of Figure 5.10d.

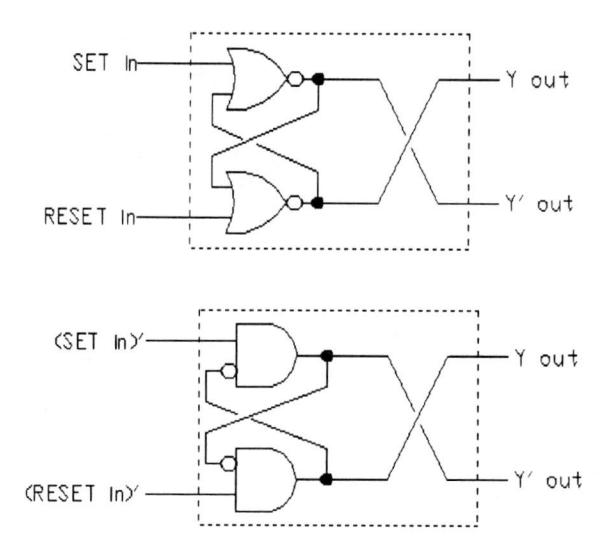

Figure 10.2 Logical Bistable Elements.

An illustration of a single component fluidic bistable device is shown in Figure 10.3. It should be noted that the two ports marked *control* are the steering ports and could be labeled SET and RESET. The *source* directs the flow through the component. The device shown is a wall attachment element. It is thought that this component functions due to the fact that there is a lower pressure zone next to the wall than exists on the opposite side of

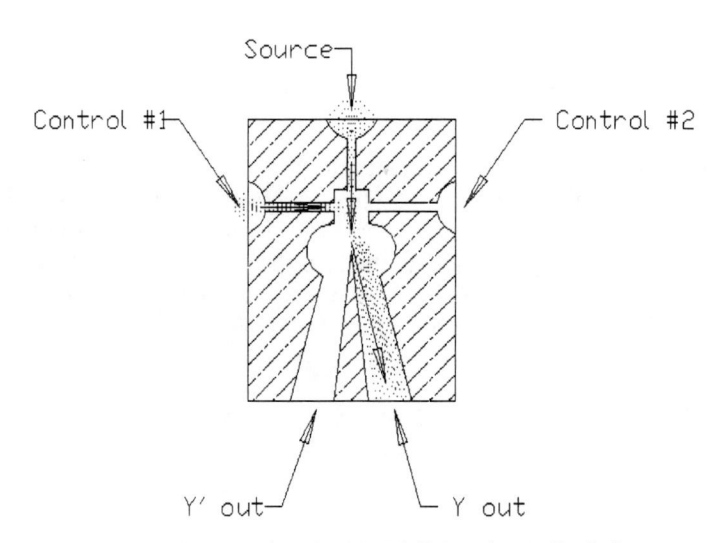

Figure 10.3 Fluidic Bistable (Wall Attachment Device).

Sec. 10.2   Active Memory Design Techniques   161

the jet (Coanda effect), but the mechanics of this have not been completely explained.

Consider that a flow is always directed from the source toward the output ports or *receivers* in an active fluidic element such as the one illustrated. If a flow is passed through the port called *control #1*, the main jet will be deflected to the output port marked *Y out* through the mechanism of momentum interchange. When this control flow ceases, the main jet will remain in the deflected position due to Coanda effect mentioned above as long as the main flow continues and no flow emanates from *control #2*. Of course, when flow passes through *control #2*, the main jet will deflect so that flow will appear at *Y' out*, and will remain in this position.

10.2 ACTIVE MEMORY DESIGN TECHNIQUES

Consider the Logic Specification Chart of Chapter 7, Figure 7.10, and its derived Reduced Specification Chart of Figure 7.15 which is reproduced below as Figure 10.4. While all automation design and analysis starts from a problem statement in the form of a PFT or LSC, they should be implemented from the RSC. The uniqueness of the active MEMORY (classical) analysis is that it carries the problem far from the RSC into a much reduced, and hopefully optimal, condition.

OUTPUT STATES $z_1 z_2$		NEXT STATES INPUT STATE: In			
	n:	1	2	3	4
	a	0	1	1	0
	b	0	0	1	1
0 0		[1]	3	7	4
1 1		1	–	6	[2]
0 1		1	[3]	6	2
0 1		1	3	5	[4]
1 1		1	3	[5]	2
1 0		1	3	[6]	4
1 1		1	3	[7]	4

Figure 10.4 Reduced Specification Chart.

10.2.1 First Steps

The first step in the classical analysis, after having derived the RSC, is to reduce the RSC to its most reduced form, called (naturally enough) the Most Reduced Specification Chart, or MRSC. This is done by first eliminating all references to the output states (they will be reinserted later), then combining *compatible* states into single rows. The first step is obvious; however *compatibility* should be defined prior to proceeding.

Two rows are *compatible* iff each component of the row contains either the same state (either stable or transitional), or don't-care designations. For example, the first and fourth rows of the RSC of Figure 10.4 are *not* compatible since they differ in the case of I_3 where row 1 contains the transitional state 7 and row 4 contains 5. Row 1 *is* compatible with row 7 however, since all the input states result in the same next state, either stable or transitional.

In order to implement this analysis, a pair-by-pair comparison of *all* the rows is done. One technique of accomplishing this requires *temporarily* naming the rows using the numbers of the stable states and then developing an Equivalent Pairs Chart, similar to that shown in Figure 7.11. The example chart is based on the RSC of Figure 10.4, and is illustrated in Figure 10.5.

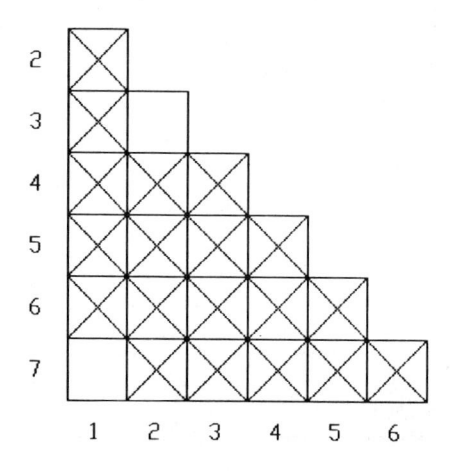

Figure 10.5 Equivalent Pairs Chart for System of Figure 10.4.

Once the Pairs Chart is completed, a Merging Graph such as that shown in Figure 7.12 can be made, if necessary. Due to the trivial nature of the case under consideration such a graph is unnecessary, and a State Table similar to that in Figure 7.13 can be made. The State Table for this problem is shown in Figure 10.6. It can be readily seen that all the minimal sets are essential. The MRSC can be generated using the information from the RSC and the State Table. Figure 10.7 shows the MRSC for the system under study.

Sec. 10.2 Active Memory Design Techniques

As stated before, this system is trivial and does not really represent one with more typical complexities. A less frivolous system is shown in the RSC of Figure 10.8; its EPC is given in Figure 10.9, its Merging Graph in Figure 10.10, its State Table in Figure 10.11, and the resulting MRSC in Figure 10.12. A discussion follows.

Maximal Sets	Sets						
	1	2	3	4	5	6	7
* (1,7)	[X]						[X]
* (2,3)		[X]	[X]				
* 4				[X]			
* 5					[X]		
* 6						[X]	

Figure 10.6 State Table. Data from Figure 10.5.

NEXT STATE
Input State
$I[1]$ $I[2]$ $I[3]$ $I[4]$

	$I[1]$	$I[2]$	$I[3]$	$I[4]$
a	0	1	1	0
b	0	0	1	1
	[1]	3	[7]	4
	1	[3]	6	[2]
	1	3	5	[4]
	1	3	[5]	2
	1	3	[6]	4

Figure 10.7 Most Reduced Specification Chart (MRSC). Data from Figure 10.6.

OUTPUT STATES		NEXT STATES INPUT STATE: In			
$Z_1 Z_2$	n:	1	2	3	4
	a	0	1	1	0
	b	0	0	1	1
0 0		[1]	3	–	2
0 1		–	–	4	[2]
1 1		5	[3]	6	–
1 0		–	3	[4]	–
1 0		[5]	–	–	7
1 1		–	9	[6]	8
0 1		–	–	6	[7]
0 0		–	–	10	[8]
0 0		11	[9]	–	–
0 1		–	12	[10]	–
1 1		[11]	–	–	8
1 0		1	[12]	–	–

Figure 10.8 Reduced Specification Chart (RSC).

A pair-by-pair analysis of the rows of the RSC is made, and the results posted in a State Table such as the one shown in Figure 10.9. It is strongly suggested that the reader perform this operation to verify the table's accuracy. Once this is done, a Merging Graph can be constructed, as has been done in Figure 10.10

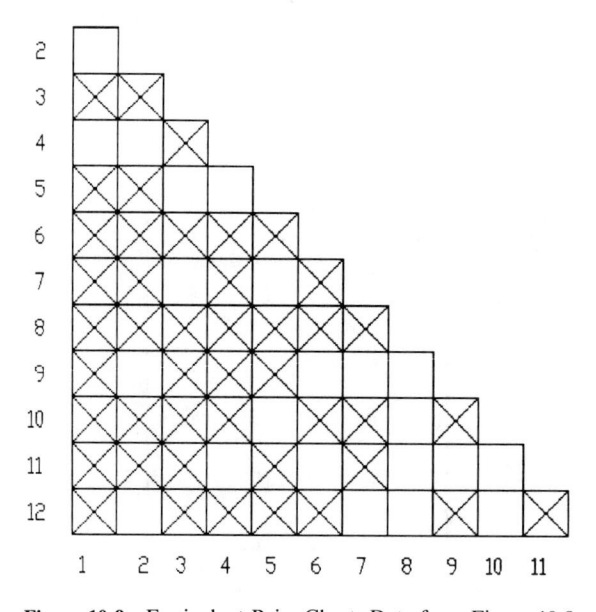

Figure 10.9 Equivalent Pairs Chart. Data from Figure 10.8.

Sec. 10.2   Active Memory Design Techniques

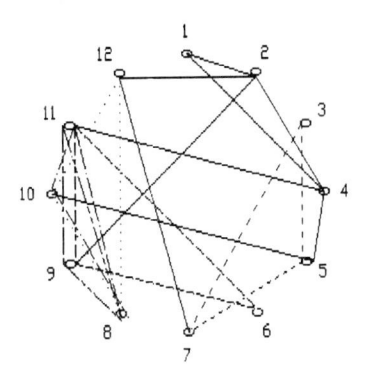

Figure 10.10 Merging Graph. Data from Figure 10.9.

Exactly the same procedure is used for this Merging Graph as was outlined in Section 7.5 and Figure 7.12. The Minimal Sets are identified, and a State Table is developed in the same manner as was shown in Chapter 7. Consider the Table of Figure 10.11.

Maximal Sets	Sets											
	1	2	3	4	5	6	7	8	9	10	11	12
* (1,2,4)	[X]	(X)		(X)								
* (3,5,7)			[X]		(X)		[X]					
* (6,9,11)						[X]			(X)		(X)	
A (8,9,11)								X	(X)		(X)	
B (8,10,11)								X		X	(X)	
C (8,10,12)								X		X		X
D (2,9)		(X)							(X)			
E (2,12)		(X)										X
F (4,5)				(X)	(X)							
G (4,11)				(X)							(X)	
H (5,10)					(X)					X		
I (7,12)							(X)					X

Figure 10.11 State Table. Data from Figure 10.10.

Upon completing the table by inserting Xs at the appropriate junctions of Minimal Set rows and Set columns, it is evident that the first three Minimal Sets are essential, since sets contained in them appear nowhere else in the table. They are therefore *square bracketed* as shown. Within these Minimal Sets, sets that are not unique are parenthesized. The other Minimal Sets containing the same sets as those that have been square bracketed or parenthesized are then identified through the use of brackets as illustrated. Once this is completed, it can be seen that the only sets *not* covered are numbers 8, 10, and 12. Since there is a Minimal Set that contains these sets (marked C on the table), it is selected as the *Chosen Essential Set*. It should be mentioned here that while the visual examination of the State Table will ultimately lead to the selection of appropriate Chosen Sets, an algebraic technique is available.

Once again, consider all the non-essential Minimal Sets (those marked *A* through *I*, inclusive). A conjunctive statement can be developed to express the value of the chosen set X_c as the ANDing of disjunctive statements from the uncovered rows, that is: $(A + B + C)$ from Set 8, $(B + C + H)$ from Set 10, and $(C + E + I)$ from Set 12, then

$$X_c = (A + B + C)(B + C + H)(C + E + I)$$

which can be expanded and then reduced to

$$X_c = C + BE + BI + AEH + AHI$$

From this expression, it can be seen that the Chosen Set(s) can be one or any combination of the terms of the above disjunctive expression. The rule to follow is to select the simplest term that will satisfy the expression. In this case, the single variable term *C* is chosen. This Minimal Set contains sets 8, 10, and 12. It is the same selection that was made by the inspection process. All things being equal, this is to be expected.

	NEXT STATE			
	Input State			
	I[1]	I[2]	I[3]	I[4]
a	0	1	1	0
b	0	0	1	1
	[1]	3	[4]	[2]
	[5]	[3]	6	[7]
	[11]	[9]	[6]	8
	11	[12]	[10]	[8]

Figure 10.12 Most Reduced Specification Chart (MRSC). Data from Figure 10.11.

10.2.2 Operational Flow Charts

The next phase of the active MEMORY design process has two steps. The first is to determine the number of active MEMORY elements required

Sec. 10.2 Active Memory Design Techniques

to satisfy the demands of the automation system. The second is to sequence these memories to avoid *races*, which are the principal sequential hazards.

As an example, consider the system shown in the MRSC of Figure 10.7. In order to differentiate among the five rows that appear in this chart, at least n memories will be required, where

$$2^n \geq \text{number of rows}$$

It is evident that at least three MEMORY elements will be required for this system. Consider what happens if the three MEMORY assignments are made casually by arbitrarily coding them in a Left Gray Code. This assignment scheme is shown in the Operational Flow Chart of Figure 10.13. Note that each of the rows has been identified by letter A through H.

	MEMORY $Y_{[1]}\ Y_{[2]}\ Y_{[3]}$	NEXT STATE Input State $I_{[1]}\ I_{[2]}\ I_{[3]}\ I_{[4]}$ a 0 1 1 0 b 0 0 1 1
A	0 0 0	[1] 3 [7] 4
B	1 0 0	1 [3] 6 [2]
C	1 1 0	1 3 5 [4]
D	0 1 0	1 3 [5] 2
E	0 1 1	1 3 [6] 4
F	1 1 1	— — — —
G	1 0 1	— — — —
H	0 0 1	— — — —

Figure 10.13 Initial Operational Flow Chart (OFC).

Assume that the automation system under consideration is in its *start-up* state, where the inputs $abY_{[1]}Y_{[2]}Y_{[3]} = 00000$. Since it is extremely unlikely (although certainly possible) that more than one input would change at any given instant, only *one* state change at a time is permitted in any system analysis. The term used to describe simultaneous or near simultaneous changes of state is a *race*.

In order to shift the system to state [2], the inputs must first change to state 00010 to move from row A to row D, then change to 01010 to a transitional state 2, or change from 00000 to 00100 to 01100 to stable state [2]. Notice that only one state change at a time has been allowed. Now consider that the first input change sequence (00000 to 00010 to 01010) occurred to transitional state 2. In order to move to stable state [2], the inputs would have to change from 01010 to 01100, BUT THIS IS A MULTIPLE STATE CHANGE, which is not permitted. Therefore, the inputs must change either from 01010 to 01110 to 01100, or from 01010 to 01000 to 01100.

In the first sequence of state changes, the system finds itself in stable state [4] rather than [2]. In the second series, it goes from transitional state 2 to transitional state 4. In the first case, states [2] and [3] have been totally missed resulting in a potential disaster. In the second, one finds the system *bouncing* between two transitional states without really knowing what to expect. Both conditions could be catastrophic!

In addition to the race described above, others exist in the system as shown. Consider the transition from $I_{[4]}$-E to $I_{[4]}$-C and from $I_{[3]}$-B to $I_{[3]}$-E (where *three* transitions are required). Obviously, the assignment of MEMORY states cannot be done capriciously!

Unfortunately, the selection of chosen secondary adjacencies and the assignment of MEMORY states is an intuitive process that can be helped by mnemonic tools, but ultimately becomes an *effect* to the *cause* of the whims of the designer. For this reason, it is reasonable to state that there are as many *correct* answers to a given automation design problem as there are individuals solving the problem.

The tool of choice in the assignment of MEMORY states is the adjacency graph coupled to a Karnaugh Map. The technique is to analyze the system, row by row, to establish adjacencies (single state-change transitions), both primary and secondary.

Definitions are once again in order. A *primary adjacency* requirement is one where there is one and only one transitional state to its corresponding stable state within a given input state configuration. A *secondary adjacency* requirement is one where there is more than one transitional state associated with a stable state.

Consider row A in Figure 10.13. For input state $I_{[1]}$ there is a secondary adjacency requirement to rows B through E for state 1, for state $I_{[2]}$ a similar condition exists in the case of state 3. For $I_{[3]}$ there is no adjacency requirement since there are no transitional states. In the case of state $I_{[4]}$, the secondary adjacency requirement involves only rows A, C, and E for state 4. This situation is illustrated in Figure 10.14. Notice that:

- Adjacencies are shown as a line connecting the row under consideration to its *adjacent* rows
- Each line is identified by the number of the state involved
- Primary adjacencies are indicated by circling the appropriate row name
- Other adjacencies are indicated by marking the appropriate row name with an X

Sec. 10.2   Active Memory Design Techniques   169

- All don't-care entries are by definition potential secondary adjacencies that are indicated on the chart by putting the set letter identification in parentheses.
- If there are *n* MEMORIES involved, there can be no more than *n* adjacencies to any given set.

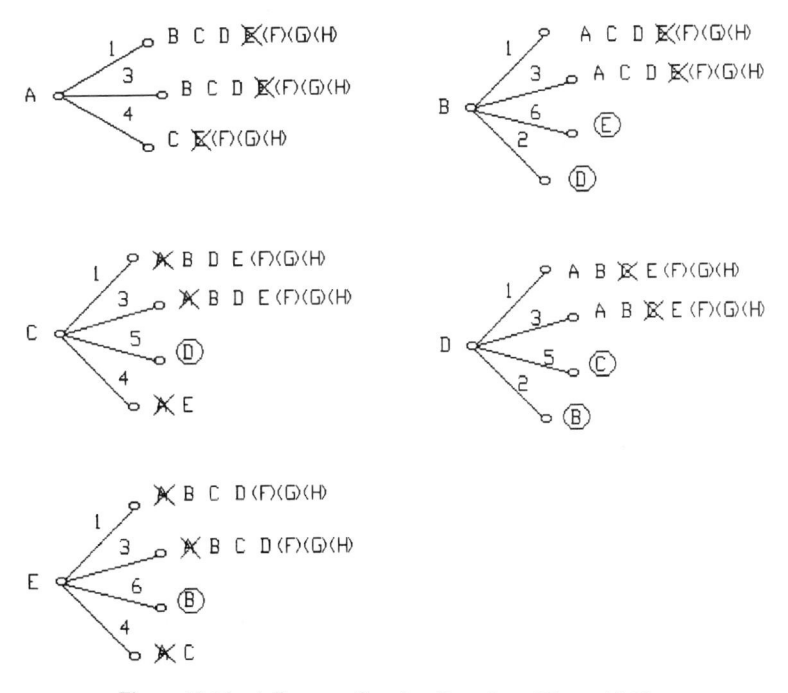

Figure 10.14 Adjacency Graphs. Data from Figure 10.13.

On studying Figure 10.14, it can be seen that states *D* and *E* must be primarily adjacent to state *B* (and vice versa, of course), as must states *C* and *D*. Since these states vary only in their memory assignments, they can be differentiated among by changes in the MEMORY assignment only. This is done by creating a Karnaugh Map where the inputs are the MEMORY states and the contents are the state names.

State *B* is the basic element, and can be *correctly* inserted into any one of the eight cells in the map. State *D* requires the first primary adjacency, but is limited to any one of three locations. Then state *E* is assigned one of two locations. State *C* is assigned. So far there are 48 different correct mappings that could be done. All the primary adjacencies have been considered, the only set still unassigned is *A*. Since *A* can have an adjacency with *B*, *C*, or *E*, it could be assigned to any of the remaining four vacant cells in the map. Unfortunately, it is noticed that either the *A-E* or the *B-E* adjacency requirement can be met, but not both. In order to eliminate this problem, one of the unused MEMORY states is assigned to state *E* as shown. This assignment carries all the members of state *E*, but only in transitional form. Again, it must be pointed out that there are as many potentially correct solutions as there are individuals solving the problem. So much for a single uniquely correct answer to an automation problem. It is strongly suggested that the reader follow the various assignment possibilities so that they are convinced as to the accuracy of various solutions.

Finally, an Operational Flow Chart can be designed by using the information from the Most Reduced Specification Chart of Figure 10.7 and the Karnaugh Adjacency Map of Figure 10.15. This chart is shown in Figure 10.16. Note that the state letter designation has been retained only for identification purposes when assigning MEMORY states. It serves no other purpose, and is usually not shown.

	Y[1] Y[2]			
	00	10	11	01
Y[3] = 0	A	C		E
Y[3] = 1	B	D		E

Figure 10.15 Karnaugh Adjacency Map. Data from Figure 10.14.

MEMORY			NEXT STATE			
			Input State			
			I[1]	I[2]	I[3]	I[4]
Y[1]	Y[2]	Y[3]	a: 0 b: 0	1 0	1 1	0 1
0	0	0	[1]	3	[7]	4
0	0	1	1	[3]	6	[2]
0	1	1	1	3	[5]	2
0	1	0	1	3	5	[4]
1	1	0	–	–	–	–
1	1	1	–	–	–	–
1	0	1	1	3	6	–
1	0	0	1	3	[6]	4

Figure 10.16 Operational Flow Chart (OFC). Data from Figures 10.12 and 10.14.

Sec. 10.2   Active Memory Design Techniques   171

In the previous example, excess MEMORY capacity was available (three MEMORIES, five states; $2^3 - 5 = 3$). A significantly different situation exists when the amount of excess MEMORY capacity is minimal. Consider the MRSC presented in Figure 10.17.

	NEXT STATE			
	Input State			
	I[1]	I[2]	I[3]	I[4]
	a 0	1	1	0
	b 0	0	1	1
A	[1]	3	[2]	4
B	[5]	6	–	[7]
C	8	[9]	2	7
D	8	[6]	10	11
E	[12]	9	[13]	[4]
F	12	[3]	2	–
G	[14]	6	[10]	[15]
H	[8]	3	13	[11]

Figure 10.17 Most Reduced Specification Chart (MRSC).

The initial MEMORY requirement calculation shows:

$$2^n \geq 8, n = 3$$

The adjacency graphs for this system are developed in the same manner as before, except for the fact that adjacencies other than primary are indicated. These graphs are illustrated in Figure 10.18.

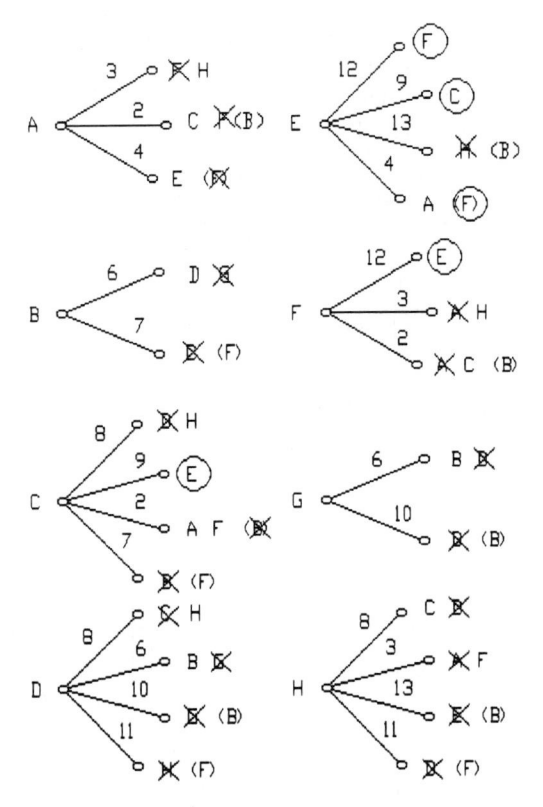

Figure 10.18. Adjacency Graphs. Data from Figure 10.17.

The first step in drawing these graphs is to identify the adjacencies and their associated paths, and to lay them out in a manner similar to that shown in the figure. Note that at this point in the process, no comments should appear on the graphs.

Next, the primary adjacencies are identified and marked on the graphs. Each one of these graphs will be identified by the letter naming the common junction. The critical paths are path C-9-E on graphs C and E, and E-12-F on graphs E and F.

It should be noticed at this point in the graphing, that for three MEMORY elements, no more than three primary adjacencies can exist. Since there are four paths in graphs C, D, E, and H, something will have to be done to eliminate at least one of each of these, or more than three MEMORIES will be required for this system.

When first checking graph E it can be seen that since an E-12-F adjacency is essential, it could also exist (if chosen) on path 4, so F on path 4 is circled to indicate a primary E-4-F adjacency, thereby reducing the number of paths to 3.

Sec. 10.2 Active Memory Design Techniques   173

On graph *G*, the two paths on 6 and 10 can both be directed to *D* so these are chosen and marked in two places with an *X*, as are the *D*-6-*G* and *D*-10-*G* paths on the *D* chart. In order to eliminate a path on the *C* graph, paths *C*-7-*B* and *C*-2-*B* are chosen. At this point, both don't-care paths have been assigned, so any further reference to these two other than paths 2 and 4 must be ignored. This forces the assignment of paths *A*-3-*F*, *A*-2-*F*, and *A*-4-*F* on graph *A*, *B*-7-*C* on graph *B*, *D*-11-*H* on graph *D*, and *H*-13-*E* and *H*-11-*D* on graph *H*. Paths *F*-3-*A* and *F*-2-*A* on graph *F* are also chosen since they were used on graph *A*, and so on. When these assignments are completed, the information is used to assign the actual adjacencies on Karnaugh Adjacency Map shown in Figure 10.19. Once again, the decisions that are made in the selection of secondary adjacencies and map cell assignments are intuitive.

$Y[1]\ Y[2]$

	00	01	11	10
$Y[3]$ = 0	B	F	E	C
$Y[3]$ = 1	G	A	H	D

Figure 10.19 Karnaugh Adjacency Map. Data from Figure 10.18.

The final Operational Flow Chart can now be generated from the data of the system shown on Figure 10.17 and the information from the map. This is illustrated in Figure 10.20. The reader should verify the adjacencies on this map, and confirm the fact that all races have been avoided.

MEMORY $Y[1]\ Y[2]\ Y[3]$	NEXT STATE
	Input State $I[1]\ \ I[2]\ \ I[3]\ \ I[4]$ a 0 1 1 0 b 0 0 1 1
0 0 0	[5] 6 2 [7]
0 0 1	[14] 6 [10] [15]
0 1 1	[1] 3 [2] 4
0 1 0	12 [3] 2 4
1 1 0	[12] 9 [13] [4]
1 1 1	[8] 3 13 [11]
1 0 1	8 [6] 10 11
1 0 0	8 [9] 2 7

Figure 10.20 Operational Flow Chart (OFC). Data from Figures 10.17 and 10.18.

10.2.3 Excitation Charts

The purpose of the Operational Flow Chart is to provide unique inputs, devoid of races, to sequential automation systems. In this section, the design process is carried almost to completion, in that the secondary state (designated as the MEMORY state) equations are developed that provide the required sequence of operation. This is the goal of any sequential design scheme.

The technique is one of developing a table called an *Excitation Chart*, and then creating *Excitation Maps* suitable for analysis. Consider first the Operational Flow Chart of Figure 10.16. The Excitation Chart of Figure 10.21 has been developed by applying the following simple steps:

- For all the stable states in any given row, place the MEMORY values into the stable state cells; and
- For the balance of the next state entries, insert the pertinent goal MEMORY state into the subject cell.

MEMORY			NEXT STATE			
			Input State			
Y[1]	Y[2]	Y[3]	I[1] a 0 b 0	I[2] 1 0	I[3] 1 1	I[4] 0 1
0	0	0	[000]	001	[000]	010
0	0	1	000	[001]	101	[001]
0	1	1	001	001	[011]	001
0	1	0	000	011	011	[010]
1	1	0	–	–	–	–
1	1	1	–	–	–	–
1	0	1	001	001	100	–
1	0	0	101	101	[100]	000

Figure 10.21 Excitation Chart. Data from Figure 10.16.

For example, in the first row, states 1 and 7 are stable states, so the value of the MEMORY state of the first row [000] is inserted into these cells. State 3 is not a stable state. The adjacent cell lies in the second row. This MEMORY value, namely 001 is assigned to the second cell in the first row of the chart. State 4 is also not a stable state, and is adjacent to the cell in the fourth row. Its next state MEMORY value 010 is placed in the appropriate cell. This process is continued until the entire Excitation Chart is

Sec. 10.2   Active Memory Design Techniques   175

completed, and all cells filled. It should be noted that the Excitation Chart is nothing more than a transformation of the Operational Flow Chart.

Once the Excitation Chart is complete, the Excitation Maps can be drawn. Examples of these are given in Figure 10.22. Although a single map could be drawn showing all the next states, it is much more convenient to create individual maps in order to minimize operator error when extracting information. Particular attention should be paid to the layout of the columns of these maps. Notice that each map is arranged such that the output variable map is grouped into halves where the left half of the map represents the 0 state of the appropriate variable, and the right half shows the 1 state. These maps serve two functions: they define the next state values of the secondary variables in the event that implementation by means of only logical operators is desired, and enable the defining of the SET and RESET equations when implementing the system using MEMORY devices.

Y[1] Y[2] Y[3]

ab \	000	001	011	010	100	101	111	110
00	0	0	0	0	1	0	–	–
01	0	0	0	0	0	–	–	–
11	0	1	0	0	1	1	–	–
10	0	0	0	0	1	0	–	–

Y[1]

Y[1] Y[2] Y[3]

ab \	000	001	101	100	010	011	111	110
00	0	0	0	0	0	0	–	–
01	1	0	–	0	1	0	–	–
11	0	0	0	0	1	1	–	–
10	0	0	0	0	1	0	–	–

Y[2]

Y[1] Y[2] Y[3]

ab \	000	010	110	100	001	011	111	101
00	0	0	–	1	0	1	–	1
01	0	0	–	0	1	0	–	–
11	0	1	–	0	1	1	–	0
10	1	1	–	1	1	1	–	1

Y[3]

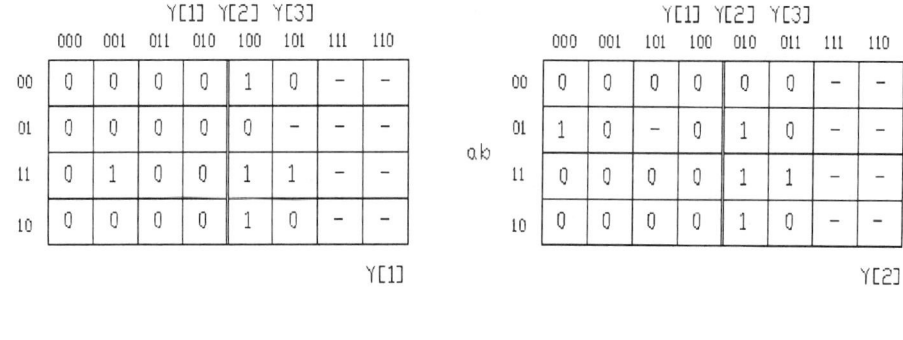

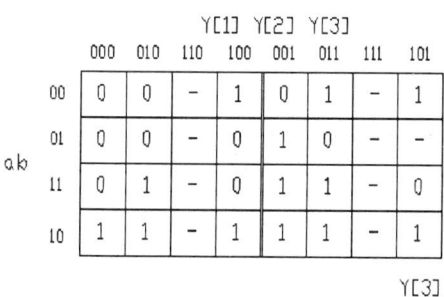

Figure 10.22   Excitation Maps. Data from Figure 10.21.

Using the techniques presented earlier, the following secondary state equations can be extracted from the maps

$$Y_1 = b'Y_1Y_3' + abY_2'Y_3 + aY_1Y_3' + bY_1Y_3 + abY_1$$

$$Y_2 = a'bY_1'Y_3' + aY_2Y_3' + abY_2 + bY_2Y_3'$$

$$Y_3 = b'Y_2Y_3 + b'Y_1 + ab' + aY_2 + bY_1'Y_2'Y_3 + a'bY_2'Y_3 + aY_1'Y_3$$

Particular attention should be paid to the fact that all the above statements are *hazard free*. The minimal form of these equations can be obtained by eliminating the last two terms of the Y_1 equation, the last term of the Y_2 equation, and the last two terms of the Y_3 equation. The use of the hazard free form is absolutely essential when implementing without the use of MEMORY elements in order to eliminate any possibility of a misdirected output.

When implementing the secondary output using MEMORY devices, the appropriate SET and RESET equations must be established, as was done when using passive MEMORY techniques. It is here that the arrangement of the maps discussed above is important. Consider the first map for the next state of variable Y_1. Observe that when the present value of Y_1 is 0, the next value of Y_1 is 1 only when the input combination is $abY_1Y_2Y_3 = 11001$. Also consider that when the present value of Y_1 is 1, the next value of Y_1 is 0 when the inputs are 00101, 01100, or 10101. These *out-of-synchronization* states indicate the proper times to SET and RESET the secondary system outputs.

The rules to follow for these assignments are:

- SET a MEMORY in the event that the present state is 0 and the next state is 1.
- RESET a MEMORY in the event that the present state is 1 and the next state is 0.

For example, the hazard-free secondary output SET and RESET equations for the three variables under study are:

SET $\quad Y_1 = abY_1'Y_2'Y_3$

RESET $\quad Y_1 = b'Y_1Y_3 + a'bY_1 + a'Y_1Y_3$

SET $\quad Y_2 = a'bY_1'Y_2'Y_3'$

RESET $\quad Y_2 = a'b'Y_2 + b'Y_2Y_3 + a'Y_2Y_3$

SET $\quad Y_3 = b'Y_1Y_3' + ab'Y_3' + aY_2Y_3'$

RESET $\quad Y_3 = a'b'Y_1'Y_2'Y_3 + a'bY_2Y_3 + bY_1Y_3$

This set of expressions can be made into a minimal set by deleting the third term of the RESET Y_1 expression, however one runs the same risk of

Sec. 10.2 Active Memory Design Techniques   177

misdirected output that was mentioned when discussing the secondary output state equations.

Following the scheme that was used before, consider now the Operational Flow Chart of Figure 10.20. The same procedure was followed as previously described to generate the Excitation Chart of Figure 10.23

MEMORY			NEXT STATE
			Input State
Y[1]	Y[2]	Y[3]	I[1] I[2] I[3] I[4]
			a 0 1 1 0
			b 0 0 1 1
0	0	0	[000] 001 010 [000]
0	0	1	[001] 101 [001] [001]
0	1	1	[011] 010 [011] 010
0	1	0	110 [010] 011 110
1	1	0	[110] 100 [110] [110]
1	1	1	[111] 011 110 [111]
1	0	1	111 [101] 001 111
1	0	0	101 [100] 000 000

Figure 10.23 Excitation Chart. Data from Figure 10.20.

The three Excitation Maps developed from the Excitation Chart are shown in Figure 10.24. Below are hazard-free secondary output equations for combinational logic implementation

$Y_1 = a'Y_2Y_3' + a'Y_1Y_3 + bY_1Y_2 + ab'Y_2'Y_3 + b'Y_1Y_3' + b'Y_1Y_2'$
$\quad + a'b'Y_1 + Y_1Y_2Y_3' + a'Y_1Y_2$

$Y_2 = Y_1'Y_2 + a'Y_2 + Y_2Y_3 + a'Y_1Y_3 + bY_2 + abY_1'Y_3'$

$Y_3 = Y_2'Y_3 + a'b'Y_3 + a'b'Y_1Y_2' + b'Y_1Y_3 + a'Y_1Y_3 + ab'Y_1'Y_2'$
$\quad + abY_1'Y_2 + abY_1'Y_3$

178   Sequential Logic Systems: Active Memory Technique   Chap. 10

	Y[1] Y[2] Y[3]							
ab	000	001	011	010	100	101	111	110
00	0	0	0	1	1	1	1	1
01	0	0	0	1	0	1	1	1
11	0	0	0	0	0	0	1	1
10	0	1	0	0	1	1	0	1

Y[1]

	Y[1] Y[2] Y[3]							
ab	000	001	101	100	010	011	111	110
00	0	0	1	0	1	1	1	1
01	0	0	1	0	1	1	1	1
11	1	0	0	0	1	1	1	1
10	0	0	0	0	1	1	1	0

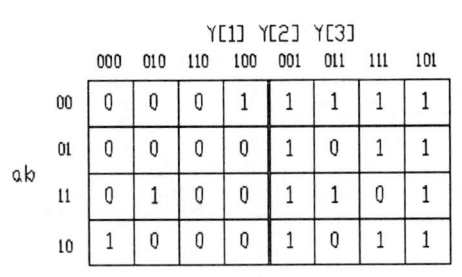

Y[2]

	Y[1] Y[2] Y[3]							
ab	000	010	110	100	001	011	111	101
00	0	0	0	1	1	1	1	1
01	0	0	0	0	1	0	1	1
11	0	1	0	0	1	1	0	1
10	1	0	0	0	1	0	1	1

Y[3]

Figure 10.24   Excitation Maps. Data from Figure 10.23.

This set can be made minimal by deleting the last four terms of the Y_1 expression and the last term of the Y_3 expression.

Below are the hazard free SET-RESET equations for the same system:

$$\text{SET } Y_1 = a'Y_1'Y_2Y_3' + ab'Y_1'Y_2'Y_3$$
$$\text{RESET } Y_1 = bY_1Y_2'Y_3' + ab'Y_1Y_2Y_3 + abY_1Y_2.$$
$$\text{SET } Y_2 = a'Y_1Y_2'Y_3 + abY_1'Y_2'Y_3'$$
$$\text{RESET } Y_2 = ab'Y_1Y_2Y_3'$$
$$\text{SET } Y_3 = a'b'Y_1Y_2'Y_3' + ab'Y_1'Y_2'Y_3' + abY_1'Y_2Y_3'$$
$$\text{RESET } Y_3 = a'bY_1'Y_2Y_3 + ab'Y_1'Y_2Y_3 + abY_1Y_2Y_3$$

This set of equations happens to be not only hazard-free; it is also minimal.

It is again strongly suggested that the reader verify all the above equations, both for checking their accuracy and for developing some expertise in the technique.

10.3 OUTPUT STATEMENTS

Formal active memory synthesis stops with the assignment of the secondary state or SET-RESET equations of a sequential automation system. Of course this does not complete the design, since the outputs have been completely ignored, and the ultimate input-output relationships remain unknown.

It is again necessary to consider the Operational Flow Chart. With it, the named states can be replaced by their output states in a manner similar to the assignment of next states when designing the Excitation Chart. Once this is accomplished, Karnaugh Output Map(s) can be generated much as the Excitation Maps were, and output relationships established.

Consider again the Operational Flow Chart of Figure 10.16 and the Reduced Specification Chart of Figure 10.4, from which it was developed. The first step of this scheme is to substitute the output values of the steady states in place of the steady state minimal set numbers. The next step is to consider all the transitional states, one at a time, on the basis of the steady states from which they can originate.

The Output Assignment Table of Figure 10.25, which describes the system under consideration is such a document. The assignment of the output values for the steady states can be seen as nothing more than the transfer of the information from the RSC to the appropriate cell in the chart. For example, state [1] has a steady state output of $Z_1Z_2 = 00$, for state [2] $Z_1Z_2 = 11$, and so on. These values are entered into the Assignment Table as shown.

MEMORY	NEXT STATE
	Input State
	I[1] I[2] I[3] I[4]
Y[1] Y[2] Y[3]	a 0 1 1 0 b 0 0 1 1
0 0 0	00 -- 11 --
0 0 1	-1 01 -1 11
0 1 1	11 11 11 11
0 1 0	01 01 01 01
1 1 0	-- -- -- --
1 1 1	-- -- -- --
1 0 1	-- -- -- --
1 0 0	10 10 10 10

$Z_1 Z_2$

Figure 10.25 Output Assignment Chart (OAC). Data from Figures 10.4 and 10.16.

The second phase of the design requires that each transitional state be considered for appropriate outputs. The scheme to follow is:

- Determine the stable states from which the transitional state can originate,
- If an output variable is the same for all of the source stable states, then it should be entered in the appropriate cell, but
- if an output variable is *not* the same for all of the source stable states, then a don't-care should be entered, *unless*
- A good case can be made for selecting a particular value for the output at this time.

Consider now transitional state 3. This state can originate from either [1] or [7]. The desired output for state [1] is 00, for state [7] 11. Since both output variables differ, the assignment of—would be perfectly appropriate. Note that this transitional state is adjacent to the related stable state, therefore a value of 01 could also be used, which might even be more judicious. But this is a decision that could probably be best made later in the assignments; so the don't-care entry is used. A similar analysis results in the assignment of output values to transitional state 4 in the first row of the chart.

In the second row, transitional state 1 can originate from either [2] or [3] where the output values are 11 and 01 respectively. The values assigned are therefore −1 as shown. Similarly, the analysis of transitional state 6 results in the same decision.

Sec. 10.4 Sequential Hazards 181

In the third row, transitional states 2 and 3 can only originate with state [5], therefore their outputs must be 11. Transitional state 1 can originate with either of the two other transitional states, so its output must also be 11, and so on.

Once the output states are all assigned, an output map can be generated. The map for this system is shown in Figure 10.26.

	Y[1] Y[2] Y[3]							
ab	000	001	011	010	100	101	111	110
00	00	-1	11	01	10	--	--	--
01	--	11	11	01	10	--	--	--
11	11	-1	11	01	10	--	--	--
10	--	01	11	01	10	--	--	--

$Z_1 Z_2$

Figure 10.26 Output Assignment Map. Data from Figure 10.25.

Based on the Assignment Map, the following output equations can be written:

$$Z_1 = Y_1 + bY_2' + Y_3$$

$$Z_2 = bY_1' + Y_2 + Y_3$$

Again, these are in hazard-free form. The system equations, that is the secondary output equations and these output equations, completely describe the specified automation system. They will be reproduced *in toto* later in order to provide a basis of comparison with other synthesis techniques.

10.4 SEQUENTIAL HAZARDS

Mention has already been made of hazards due to races and their resulting potential danger of misdirected sequencing. These hazards can be eliminated during the changes of MEMORY states by implementing either the secondary state equations or the appropriate SET-RESET equations, in hazard-free form.

In addition to this hazard, there are others, namely

- Combinational static and dynamic hazards,
- Hazards due to errors in design and/or inappropriate hardware selection, and
- Hazards resulting from races due to simultaneous, or near simultaneous, changes in the input states.

Since a rather complete discussion of combinational hazards was presented in Chapter 6, it is inappropriate that it be repeated here. In the event that the reader is unsure of some of the considerations or techniques of hazard detection and elimination, it is suggested that section be restudied.

Hazards resulting from errors in judgement can only be eliminated through practice and experience. The best advice that can be given under these conditions is "Don't make the same mistake once!"

Finally, there remains the real problem of hazards due to the racing of input signals. These can usually be eliminated through judicious selection of signal sources and timing; that is, when a multiple state change is necessary in order to correctly sequence the machine being automated, input signals should be designed so that they will not occur at nearly the same time. If this is not physically reasonable, DELAY elements will have to be inserted into the system to preclude racing. The application of DELAY elements to this type of hazard was also discussed in Chapter 6.

10.5 SUMMARY

At this point, it might be interesting to again ask the question "What is the *best* technique to use when designing logical automation systems?" As mentioned before, there is no clear-cut answer. Much depends on the nature of the specific problem. Equally important are the flexibility and preferences of the designer. For example, consider the problem defined in the Logic Specification Chart of Figure 7.10.

The first step in any sequential design is the development of a Reduced Specification Chart. In the case under consideration, this was demonstrated in the synthesis of the RSC of Figure 7.15, which was repeated as Figure 10.4.

For simple problems, or ones where many inputs are used resulting in a system with many source states, the counting and stepping methods of Chapter 8 can be profitably used. For more complex requirements, the two schemes that would most likely result in an acceptable system design are the passive and active MEMORY techniques. The Memory State Equations and Output Equations derived from the subject RSC for each are shown below for comparison and evaluation.

Memory States

PASSIVE	ACTIVE
RESET Y_1: $a'bY_3$	$Y_1(Y_3(a' + b') + a'b)$
SET Y_1: $a'bY_3' + ab'$	$abY_1'Y_2Y_3$
RESET Y_2: $a'bY_3' + ab'$	$Y_2(Y_3 + a'b')$
SET Y_2: $a'b'$	$a'bY_1'Y_2'Y_3'$
RESET Y_3: $ab' + abY_1'Y_2'$	$Y_3(a'(bY_2 + b'Y_1'Y_2') + b'Y_1)$
SET Y_3: $a'b' + abY_1$	$Y_3'(a(b' + Y_2) + b'Y_1)$

Outputs:

PASSIVE MEMORY	ACTIVE MEMORY
$Z_1 = a'bY_3' + ab(Y_2' + Y_1)$	$Z_1 = Y_1 + b(Y_2' + Y_3) + Y_2Y_3$
$Z_2 = a'b + ab' + ab(Y_1'Y_2' + Y_1Y_2)$	$Z_2 = Y_2 + Y_3 + bY_1'$

It can be seen that no clear-cut advantage lies with either procedure. This is true since there are the same number of required MEMORIES in both cases. If one strategy required fewer MEMORIES, it would probably result in a more effective design. Remember that a Classical Analysis requires considerably more engineering time and effort, resulting in significantly greater total cost.

In conclusion, it can be stated that in the event that only one automation system is to be constructed, one could afford to accept more components in order to save additional design time and costs. Where multiple systems are required, the additional engineering time is justified. The basic rule to follow is

> One pays for design once; one pays for production for the life of the product.

10.6 PROBLEMS

1. Using the RSC developed from the PFT of Figure P9.11, devise the Most Reduced Specification Chart for this system.
2. Using the RSC developed from the PFT of Figure P9.12, devise the Most Reduced Specification Chart for this system.
3. Using the RSC developed from the PFT of Figure P9.13, devise the Most Reduced Specification Chart for this system.
4. Using the RSC developed from the PFT of Figure P9.14, devise the Most Reduced Specification Chart for this system.
5. Using the RSC developed from the PFT of Figure P9.15, devise the Most Reduced Specification Chart for this system.
6. Using the RSC developed from the PFT of Figure P9.16, devise the Most Reduced Specification Chart for this system.
7. Using the MRSC developed in problem 1 above, design the Operational Flow Chart for this system.
8. Using the MRSC developed in problem 2 above, design the Operational Flow Chart for this system.
9. Using the MRSC developed in problem 3 above, design the Operational Flow Chart for this system.
10. Using the MRSC developed in problem 4 above, design the Operational Flow Chart for this system.

11. Using the MRSC developed in problem 5 above, design the Operational Flow Chart for this system.
12. Using the MRSC developed in problem 6 above, design the Operational Flow Chart for this system.
13. Using the OFC developed in problem 7 above, devise the required Excitation Chart and Maps, write the system equations, and draw a schematic design circuit for this system.
14. Using the OFC developed in problem 8 above, devise the required Excitation Chart and Maps, write the system equations, and draw a schematic design circuit for this system.
15. Using the OFC developed in problem 9 above, devise the required Excitation Chart and Maps, write the system equations, and draw a schematic design circuit for this system.
16. Using the OFC developed in problem 10 above, devise the required Excitation Chart and Maps, write the system equations, and draw a schematic design circuit for this system.
17. Using the OFC developed in problem 11 above, devise the required Excitation Chart and Maps, write the system equations, and draw a schematic design circuit for this system.
18. Using the OFC developed in problem 12 above, devise the required Excitation Chart and Maps, write the system equations, and draw a schematic design circuit for this system.

Chapter 11

AUTOMATION CASE STUDY

The following Case Study is an example of a manufacturing engineering problem that involves many phases of product design and manufacturing planning. Note that the automation system used a relatively large number of independent inputs, resulting in a very straightforward analysis using the IF-THEN technique described in Chapter 7.

This Case Study was presented at the International Conference on Hydraulics, Pneumatics, and Fluidics in Control and Automation, in Toronto, Ontario, Canada, in 1976.

11.1 INTRODUCTION

Prior to the automation of the manufacturing process for a part, a number of factors must be considered. First, adequate production quantities must be involved to warrant the expense incurred in the design and implementation of an automated system. Next, the design of the part to be produced must be analyzed as to its manufacturability by an automated process. Third, the design, construction, and testing of a machine that will generate the appropriate surfaces, as well as the control system for such an apparatus, must be provided.

11.2 INITIAL CONSIDERATIONS

In any automation system design, at least four essential questions must be addressed:

- What degree of automation is warranted by the required production quantities?
- What automation technique should be used, that is, feedback control (for continuous processes) or logical digital schemes?
- How should the automation system function?
- What automation hardware system provides the optimum combination of reliability, maintainability, rate of production, and total cost?

11.3 INITIAL ANALYSIS

The part whose production was to be considered for automation is shown in its original form in Figure 11.1b, and in its redesigned configuration in Figure 11.1c. The blank from which the part was made is illustrated in Figure 11.1a. Production quantities were estimated to be approximately 100,000 parts per year. It was predicted that this rate of demand would hold for at least five to ten years.

Sec. 11.3   Initial Analysis

Figure 11.1a Sintered Ferrite Blank.

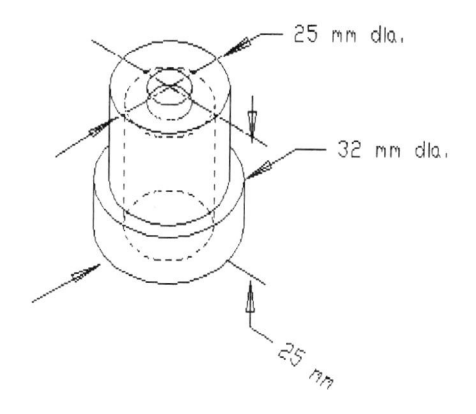

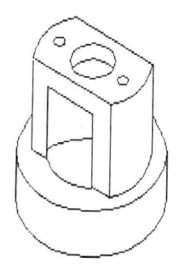

Figure 11.1b Original Part Design.

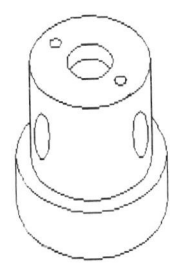

Figure 11.1c Revised Part Design.

Figure 11.1 Part Design for Case Study.

The original design was produced by a combination of milling and drilling. Milling rates were 10 pieces per hour per machine, more or less, for two machines, including both productive cutting and tool replacement times. Resharpening was done continuously on alternate sets of milling cutters.

Preliminary estimates indicated that the revised design could be produced at a rate of 90 pieces per hour if the process was fully automated, including part feeding and ejection. A rate of 40 pieces per hour could be realized if the design was changed, but the process was *not* automated. Based on these estimates, the decision was made to investigate the design of a special purpose fully automatic drilling machine using the sintered ferrite blank.

The part in question will be produced on an automated drilling machine. Since this is the case, no advantage could be realized from the use of feedback control. Logical digital automation is appropriate, especially since broken tool detection was desired. This decision was confirmed by the fact that the apparatus would best be cycled by event timing, that is by proceding to the next step after the present step is completed.

Once this decision had been made, it became apparent that the controls could be implemented using any hardware system. The possible strategies considered were electronic, electromechanical, fluidic, and valve control.

Any of these techniques could provide an automation system that would respond to an input signal in at least an order of magnitude less time than the machine response time. Therefore, speed of operation was not regarded as an important consideration. The possibility of miniaturization using electronic or fluidic components had no special advantage. Susceptibility to noise resulting in spurious operation mitigated against the use of the low-level input schemes as well. Differences in acquisition prices of the various components did not contribute any significant cost savings due to the *one-off* nature of the system. This was true due to the relatively small contribution to total cost by the components compared with that of design, implementation, and testing.

Based on the above, the selection of a hardware system was based on reliability and maintainability considerations. The production and engineering facilities at the client's factory were such that, while electronic service and maintenance capabilities were available, they could only be obtained through the engineering department (rather than production or plant maintenance). This rendered an electronic component system unfeasible since a minor failure of the system could result in excessive downtime while maintenance was being provided. Similarly, fluidic techniques required a degree of sophistication and knowledge that was totally unavailable.

Both electro-mechanical and moving-part pneumatic systems were felt to be equivalent as to ease of maintenance, but the pneumatic system had the distinct advantage of direct compatibility to both the sensing and the power output systems (see below), as well as significantly higher reliability. Either hardware technology could be modeled using two-input logic elements. Therefore, the decision was made to design the system for implementation with two-input elements, and make the final choice as to implementation after the design had been completed.

11.4 PROBLEM STATEMENT

The problem was stated as:

> Design a fully automatic drilling machine to produce the part illustrated in Figure 11.1c. Provide the required sensing and logic to enable the machine to carry out its mission.

The machine to perform the desired drilling operations was designed and is illustrated schematically in Figure 11.2. For simplicity and clarity, only one of the four side-hole drilling units (with its sensors) is shown.

Sec. 11.4   Problem Statement

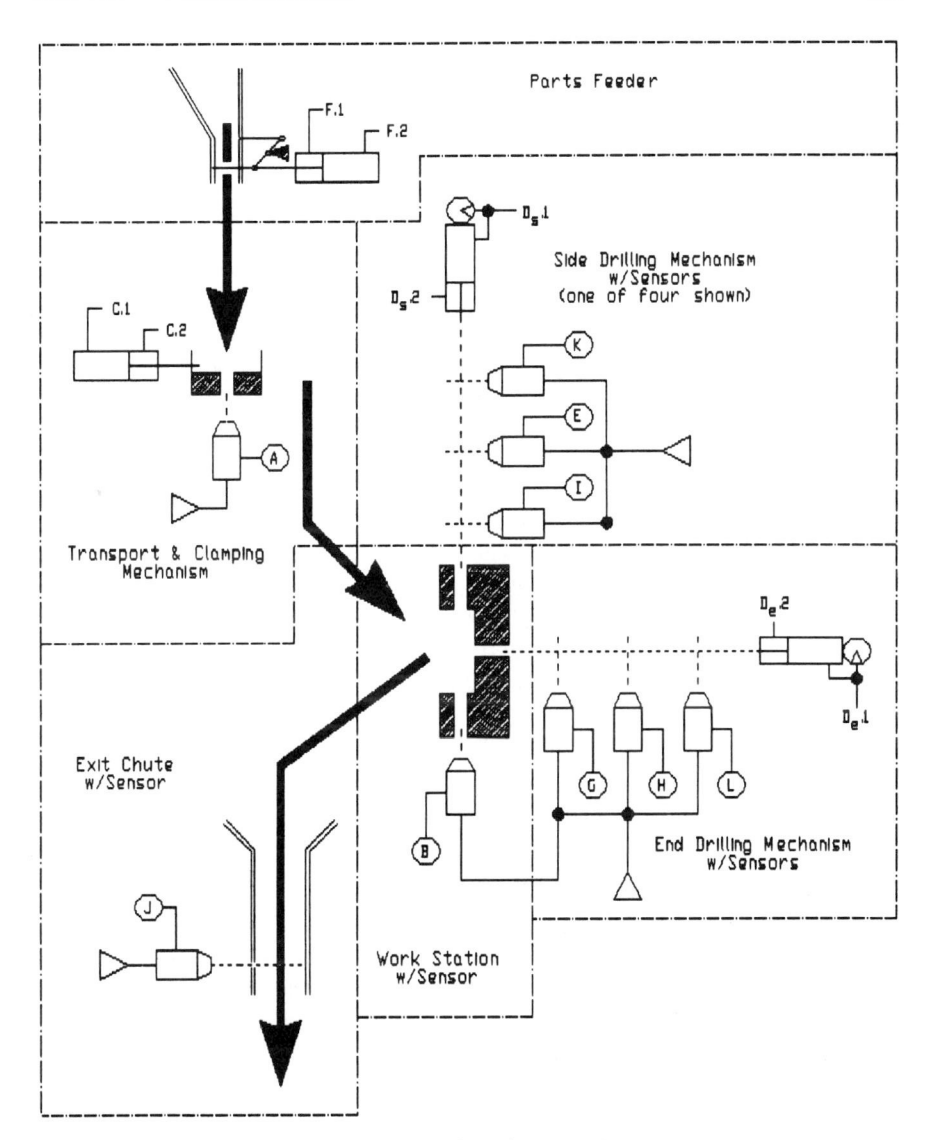

Figure 11.2 Schematic Illustration of Automatic Drilling Machine.

In the diagram, the heavy solid lines signify the flow of raw material through the machine. Sintered blanks were dumped in bulk into a vibratory feeder where they were aligned into a uniform orientation and fed by gravity into the parts feeder shown in the upper left of the schematic. One part at a time was fed to the clamping mechanism stage by the action of a cylinder whose ports are indicated by *F.1* and *F.2*. The blank was then transported to the work station and clamped in position by the cylinder whose inputs are labeled *C.1* and *C.2*. Pneumatic drilling units designated as D_s for the four side holes, and one unit named D_e for the two small end holes and the large hole which were drilled simultaneously, were selected to perform the actual drilling. Their ports were called 1 and 2 as in the other cases. The receiving station was located directly to the right of the clamping cylinder, while the work station was to the right of, and below, the receiving station.

External sensing consisted of pneumatic back pressure devices labeled *A* which detected a part in the receiving station, *B* which sensed a part in the work station, *E* and *H* which noted that the drilling units were in the back or *ready* state, *G* and *I* which detected the completion of the drilling cycles, *J* which verified that the part had cleared the machine, and *K* and *L* which acted as broken tool detectors. A separate three-position valve enabled the *reset* function in one setting, and *start* in the other. A MEMORY was set manually for power by this switch when it was operated to the *start* position, and reset by the action of any of the broken tool detectors.

Pneumatic drilling units were selected for two reasons. They were directly compatible with pneumatic control elements greatly simplifying the input-control-power system interface. In addition, the nature of the pneumatic drill unit was that it provided a constant axial force at the drill tip, rather than constant velocity as was the case with other units. This was felt to be advantageous due to the very low machinability of the sintered blanks and the concomitant tool wear and breakage that could be anticipated.

11.5 DESIGN OF THE CONTROL SYSTEM

Even though it appeared that valve logic was the way to implement the controller, it was felt that the design should be stated in the most general fashion and then implemented as optimally as feasible. To accomplish this, operational parameters were established for each step in the manufacture of the part. This statement of parameters was then converted into a Boolean expression, which in turn was incorporated into a logic diagram, from which the final system was designed.

The parameters and Boolean expressions were as follows (refer to Figure 11.2):

1. Feed the part iff the receiver is empty and the previous part has passed through the machine, or if the reset/start control is operated.

$$A'(J + R) \rightarrow (F.1)$$

Sec. 11.5 Design of the Control System

2. Reset the part feeder at some convenient time after a part has been fed.

$$R + (C.1) \rightarrow (F.2)$$

3. Move the part from the receiving station to the work station and clamp it in place iff the part is in the receiving station, and all work heads are in their respective *ready* positions, and the previous part has exited the machine.

$$AEH(F.1) \rightarrow C.1$$

4. Drill the first of the four radial holes iff the part is clamped in the work station and all other tools are retracted.

$$B(C.1) \rightarrow (D_s.1)$$

Note that this logic was repeated for the second, third, and fourth radial hole, with the additional requirement that the previous drilling unit had returned to its *ready* position.

5. Retract the first set of tools iff its drilling is completed.

$$I + R \rightarrow (D_s.2)$$

6. Drill the three axial holes iff the radial holes have all been drilled and the part is clamped in the work station.

$$EI \rightarrow (D_e.1)$$

7. Retract this last set of tools when the axial holes have been completely drilled.

$$G + R \rightarrow (D_e.2)$$

8. Unclamp and eject the part after the last set of holes has been drilled, providing that none of the broken tool detectors indicates a broken tool.

$$GH \rightarrow (C.2)$$

$$K + L \rightarrow STOP$$

The logical automation system incorporating these relationships is shown in Figure 11.3.

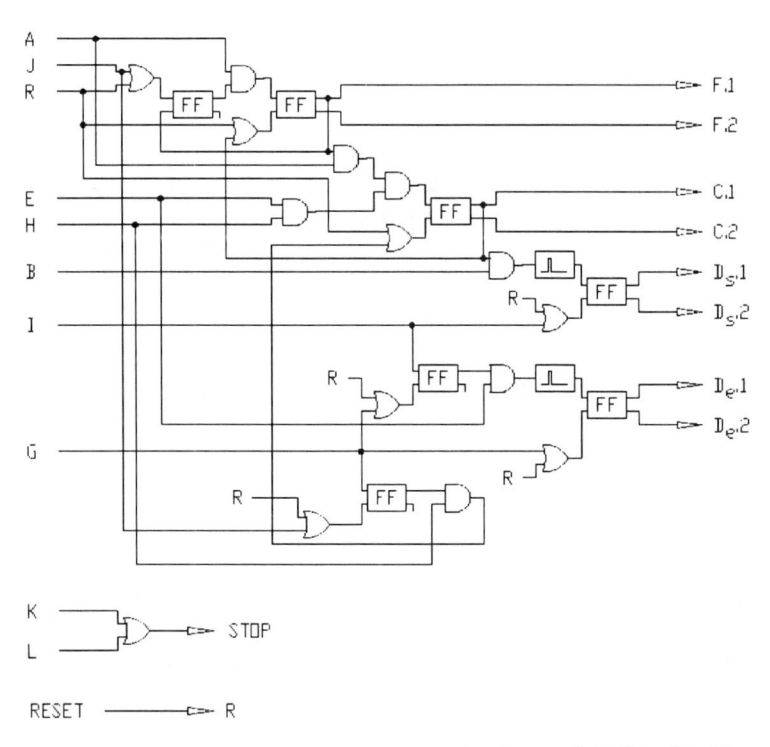

Figure 11.3 Logic Diagram of Control System for Automatic Drilling Machine.

11.6 RESULTS

The machine with its controller was constructed and put into operation. Actual results showed an overall production rate of 94 pieces per hour including down-time for drill replacement and maintenance. The total cost of the drilling machine, part feeder, and controller was slightly less that $12,000.00 at the time of construction.

In order to rationally evaluate the effectiveness of the automation, the effect of the change in design must first be considered. To do this, the estimated cost for manually drilling the six holes in the revised part was compared to the milling and drilling required for the original design.

Original Design	Revised Design (Manual Fabrication)
Labor (L_o): $1.00/pc	Labor (L_m): $0.25/pc
Tooling (T_o): 0.33/pc	Tooling (T_m): 0.01/pc
Sharpen & Set (S_o): 0.67/pc	Sharpen & Set (S_m): 0.01/pc
	Drill Jig (D_m): $350.00

Sec. 11.6 Results

The *break-even* quantity Q can be computed using the relationship

$$Q_o(L_o + T_o + S_o) = Q_m(L_m + T_m + S_m) + D_m \mid Q_o = Q_m = Q$$

resulting in a break-even point Q at 202 pieces, less than one day's production. Obviously, the design change was worthwhile, all things being equal.

Next, the costs incurred using the automatic drilling machine and manual machining, for the revised design, had to be analyzed.

Revised Design
(Automatic Machining)

Labor (L_a) : $0.11
Tooling (T_a) : 0.01
Sharpen & Set (S_a): 0.01
Machine Cost (M_a) : $12000.00

The break-even quantity was again calculated using a slight modification of the above model

$$Q_m(L_m + T_m + S_m) + D_m = Q_a(L_a + T_a + S_a) + M_a \mid Q_m = Q_a = Q$$

which resulted in a "break even" quantity of 83,214 pieces. Considering that annual usage was in excess of 100,000 parts, this meant that the payoff for the entire capital investment would occur within ten months.

The final comparison was made between the original design, original process; and the revised design, automated process, using the relationship

$$Q_o(L_o + T_o + S_o) = Q_a(L_a + T_a + S_a) + M_a \mid Q_o = Q_a = Q$$

which resulted in a break-even quantity of 6417 pieces. Considering a production rate of 94 pieces per hour, this indicates a payoff period of almost exactly ten days, even when considering a productive day of only 6.8 hours. By substituting the quantity 100,000 pieces in the above relationship, a resulting annual saving of $175,000.00 can be demonstrated.

Chapter 12
COMPUTER ASSISTED DESIGN

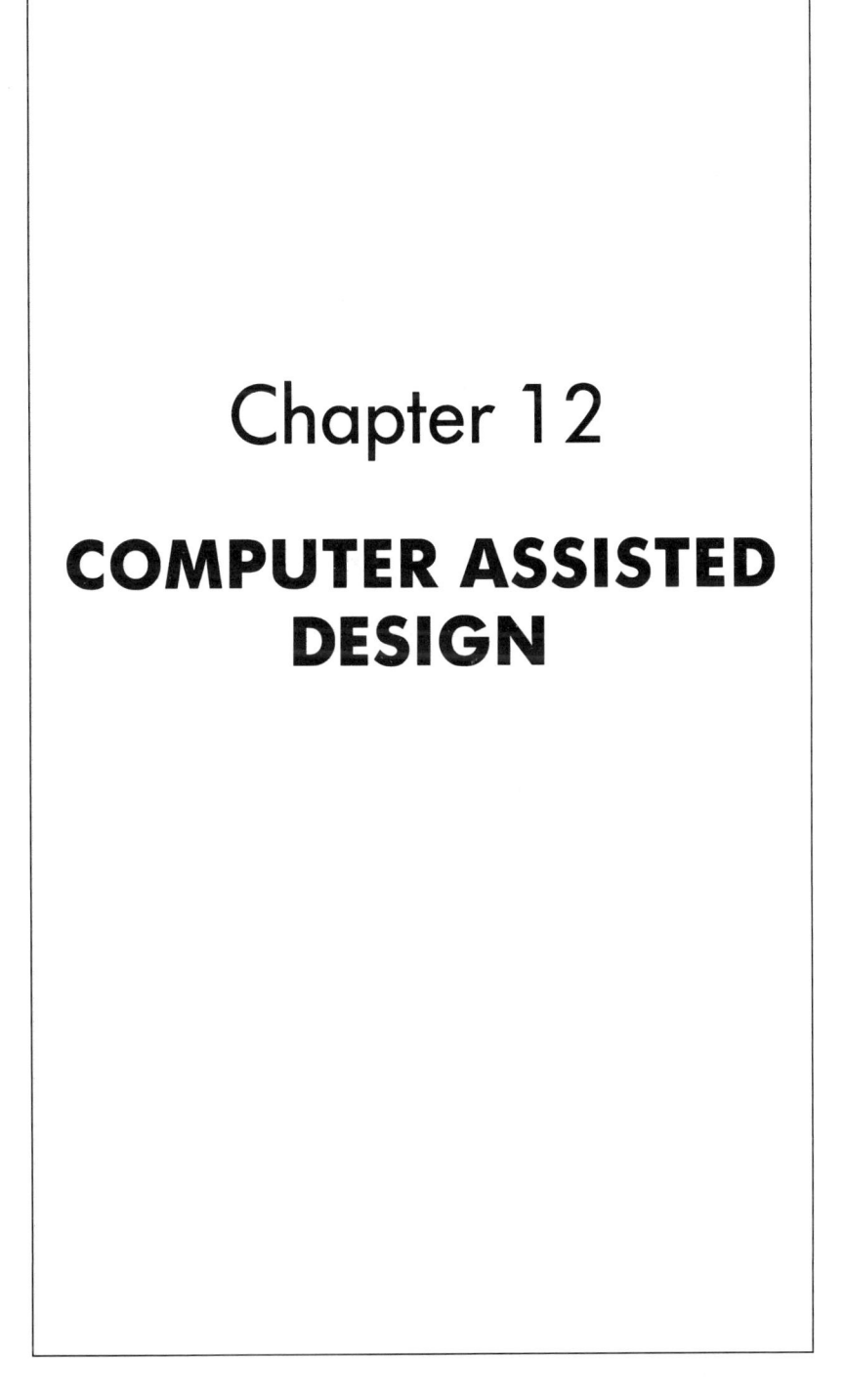

Numerous computer programs have been written to aid in the simplification and design of logical automation systems. Unfortunately, most of these fall into one or both of two categories:

- Those programs that have been written for use on main frame computers, and
- Those programs that have not been adequately tested and verified for accuracy and reliability.

With the wide distribution of desktop microcomputers, the older programs written for main frame machines are of limited value due to their cumbersome nature coupled with the lack of accessibility. The second category of programs were usually generated in an academic environment, and for one reason or another, have *glitches* which are difficult for anyone except the author to extract and correct.

For these reasons, the only programs that are mentioned in this text have been written specifically for microcomputers operating under either MS/DOS or PC/DOS operating systems. They have been in use for a number of years and have been extensively tested. Listings of these programs are given in the appendix.

12.1 PRIMP

This program simplifies logical statements by means of a prime implicant analysis. User information is presented below. The problem shown in Figure 4.4a was input as eleven individual minterms. This sample problem is documented in Figure 12.1.

Sec. 12.1 Primp   197

```
                    *** BOOLEAN MINIMIZATION PROGRAM **

                    THIS FUNCTION CONTAINS   5 VARIABLES

                    A LISTING OF THE INPUT DATA FOLLOWS
                    TRUE MINTERMS = 1
                    FALSE MINTERMS = 0
                    REDUNDANT MINTERMS (DON'T CARES) = -1

         0 =  1,     1 =  1,     2 =  1,     3 =  1,     4 =  0,
         5 =  0,     6 =  0,     7 =  1,     8 =  0,     9 =  0,
        10 =  0,    11 =  0,    12 =  0,    13 =  0,    14 =  1,
        15 =  1,    16 =  0,    17 =  0,    18 =  0,    19 =  0,
        20 =  0,    21 =  0,    22 =  1,    23 =  1,    24 =  0,
        25 =  0,    26 =  0,    27 =  0,    28 =  0,    29 =  1,
        30 =  0,    31 =  1,

    THE FOLLOWING IS A LIST OF THE PRIME IMPLICANTS OF THE
    MINIMIZED FUNCTION.

         ESSENTIAL PRIME IMPLICANTS ARE SO LABELED, AND
         PRIME IMPLICANTS SELECTED FROM A CYCLIC CHART ARE
         LABELED AS CHOSEN.

         NO.  COST       PRIME IMPLICANTS
                       A  B  C  D  E
          1    3       0  0  0  X  X   ESSENTIAL
          2    4       0  0  X  1  1
          3    3       X  X  1  1  1   CHOSEN
          4    4       0  1  1  1  X   ESSENTIAL
          5    4       1  0  1  1  X   ESSENTIAL
          6    4       1  1  1  X  1   ESSENTIAL

    X INDICATES A MISSING VARIABLE, 0 INDICATES A COMPLEMENTED
    VARIABLE AND 1 INDICATES A TRUE VARIABLE.
    THE FUNCTION IS REPRESENTED BY THE SUM OF BOTH THE
    ESSENTIAL AND THE CHOSEN PRIME IMPLICANTS.


             CONSTRAINT TABLE

         COVERED     COVERING PRIME
         MINTERM       IMPLICANTS.
            7         2    3
```

Figure 12.1 PRIMP Printout.

This microcomputer package is designed to simplify logic expressions using the prime implicant method. A working knowledge of this technique and of microcomputers is assumed.

Before the prime implicant minimization program is used, the logic statement to be simplified should be converted to binary representation. This can be seen in the following example problem:

$$Z = ABC' + AB + AB' + A'B + A$$
$$Z = 110 + 11\text{-} + 10\text{-} + 01\text{-} + 1\text{--}$$

where 1 represents the presence of a literal in a minterm, 0 represents the presence of a complemented literal in a minterm, and a - signifies that the literal is not present in a minterm.

The expression will be entered in this binary representation, as seen in the sample problem. An expanded form of the above problem is given as part of the output of the program, where 0 represents the binary number *zero*, which in turn represents the minterm 000 in this example, or $A'B'C'$. $0 = 0$ implies that $A'B'C'$ is an off term, or part of the $Z_{(off)}$ statement. The sample problem above would be expanded as follows:

110 = 110   11- = 110, 111   10- = 100, 101   01- = 010, 011
1-- = 100, 101, 110, 111

This is listed in the output as:

0 = 0, 1 = 0, 2 = 1, 3 = 1, 4 = 1, 5 = 1, 6 = 1, 7 = 1

From this it can be seen that, for instance, $4 = 100 = AB'C'$ is present in the expanded logic statement.

The minimized expression is represented in much the same way. The number of the minterm is simply which minterm it is (first, second, etc.), and the cost is a count of the number of literals in the minterm. The constraint table shows any minterms which are not essential and not chosen. By looking at the output,

NO.	COST	PRIME IMPLICANTS			
		A	B	C	
1	1	X	1	X	ESSENTIAL
2	1	1	X	X	ESSENTIAL

it can be seen that the minimized logic statement is:

$$Z = B + A$$

To run the program, first be sure the computer has been booted with DOS. Then, insert the diskette containing the program into the default drive of the computer. Note that the program will not run correctly if the diskette is not in the default drive! Now type PRIM and return. Answer the questions that appear on the screen. It is only necessary to hit return after replying to questions about the number of minterms. In all other cases, as soon as a key is struck, the answer is read by the computer. Do not use the letter *O* in place of the number zero. Also, the number pad to the right of the keyboard cannot be used unless the NumLock key is engaged.

After the input is entered, the simplified expression will be found. During this time, do not be alarmed by messages saying "Stop-Program terminated." These only mean that everything is working properly.

Sec. 12.2 Trutab   199

When answering the questions about printouts and the version of DOS being used, it is not necessary to hit the return key. Also, the first time an answer is printed, you will be asked for the name of a list device. Simply hit return for the default, which works on most computers. If this does not work, the DOS manual should be consulted for technical details.

This program creates work files on the diskette; so be sure to leave room for them.

The answer is contained in split form, one screen per file, in the files ANS.1 and ANS.2. The entire output is in the file PRTANS. These can be viewed or printed out as desired.

The program is limited to expressions of eight literals or less. The main limitation is computer memory. Most compilers limit the data segment to 64K. To expand the program, source files must be changed and recompiled.[10]

12.2 TRUTAB

This program will develop a most efficient Truth Table, that is one that shows only the smaller of the true or false tables. It has proven to be invaluable in the verifying of results of logical simplification. As an example of its use, consider the prime implicant optimization of the problem as stated in Figure 4.4.

$$Z = ABCE + AB'CD + A'BCD + A'B'C' + CDE$$

This expression was input into TRUTAB. Its user documentation is illustrated in Figure 12.2.

```
INPUT LOGIC FUNCTION AS '240 Z=fn(A, B, ... , H)' THEN <RUN 70>
Break in 60
Ok
240 Z = (A AND B AND C AND E) OR (A AND NOT B AND C AND D) OR (NOT A AND B AND C
 AND D) OR (NOT A AND NOT B AND NOT C) OR (C AND D AND E)

RUN 70
ENTER THE TRUTH VALUE OF Z: 1=TRUE, 0=FALSE, DC=DON'T CARE? DC

NO. OF VARIABLES: ? 5
```

Figure 12.2 TRUTAB Input Instructions.

This program is written in BASIC and is extremely intolerant of any violation of its commands. The output of the sample problem is shown in Figure 12.3. It should be noted that the truth table resulting from choosing the *DC* option is exactly that which was inputted in the PRIMP program, which should not be unexpected.

A	B	C	D	E	F	G	H	Z
0	0	0	0	0				1
0	0	0	0	1				1
0	0	0	1	0				1
0	0	0	1	1				1
0	0	1	1	1				1
0	1	1	1	0				1
0	1	1	1	1				1
1	0	1	1	0				1
1	0	1	1	1				1
1	1	1	0	1				1
1	1	1	1	1				1

240 Z = (A AND B AND C AND E) OR (A AND NOT B AND C AND D) OR (NOT A AND B AND C AND D) OR (NOT A AND NOT B AND NOT C) OR (C AND D AND E)

Figure 12.3 TRUTAB Output.

SELECTED BIBLIOGRAPHY

1. Bensch, L.E. *A Computer Program for Synthesis of Regularly Activated Fluid Logic Sequential Circuits.* Stillwater, OK: Oklahoma State University, 1970.
2. ———, Chan, C.C., & Surjaadmadja, J.B. *Clasyn Program and User's Guide for the Classical Synthesis of Fluid Logic Networks.* Unpublished paper, Stillwater, OK: Oklahoma State University, 1973.
3. Boole, George. *The Mathematical Analysis of Logic.* Cambridge: MacMillan, Braklay, and MacMillan, 1847; rpt. Oxford: Basil Blackwell, 1951; New York: Dover Publications, 1951.
4. Chen, P.I. & Lee, Y.H. "State Diagram Synthesis for Feedback Circuits." *Paper 73-WA/FLCS-2,* presented at the ASME Winter Annual Meeting, 1973.
5. Cheng, R.M.H. & Foster, K. "A Computer-Aided Design Method Specially Applicable to Fluidic-Pneumatic Sequential Control Circuits." *Paper 70-WA/FLCS-17,* presented at the ASME Winter Annual Meeting, 1970.
6. Cole, J.H. *Synthesis of Optimum Complex Fluid Logic Sequential Circuits.* Unpublished Ph.D. Dissertation, Stillwater, OK: Oklahoma State University, 1968.
7. ——— & Fitch, E.C. "Synthesis of Fluid Logic Control Circuits," *Proceedings of the JACC,* pp. 425ff, 1969.
8. David, R. & Richard, J.P. "Cellular Logical Networks," ASME Paper 75-Aut-D, *Journal of Dynamic Systems, Measurements, and Controls,* 1975.
9. Dietmeyer, D.L. *Logic Design of Digital Systems.* Boston, MA: Allyn & Bacon, 1971.
10. Euting, J. Unpublished research paper, Oxford, Ohio: Miami University, 1987.
11. Finkbeiner, D.T. *Introduction to Matrices and Linear Transformation.* San Francisco, CA: W.H. Freeman & Co., 1966.
12. Fitch, E.C. "Synthesis of Fluid Logic Systems," *Paper 70-WA/DE-47,* presented at the ASME Winter Annual Meeting, 1970.
13. ——— & Maroney, G.E., "The Mathematical Synthesis and Analysis of Fluid Logic Networks," *Fluidics Quarterly,* Vol.4 No. 3, July 1972.
14. ——— & Surjaadmadja, J.B. *Fluid Logic,* New York: McGraw-Hill Book Co., 1978.
15. Friedman, S.B. "Design of a Fully Automated Machine Tool Using Pneumatic Low-Cost Automation Techniques," *Proceedings of the International Conference on Hydraulics, Pneumatics, and Fluidics in Control and Automation,* Toronto, Ontario, Canada: April 1976.
16. ——— & Harvey, A.E. "A Technique for the Selection of Hardware in Automatic Control Systems." *Paper 76-811,* ISA Annual Conference, Oct. 1976.
17. Harrison, M.A. *Introduction to Switching and Automata Theory.* New York: McGraw-Hill Book Co., 1965.
18. Hill, F.J. & Patterson, G.R. *Introduction to Switching Theory and Logical Design,* 2d ed., New York: John Wiley & Sons, 1974.
19. Huffman, D.A. "The Synthesis of Sequential Switching Circuits," *Journal of the Franklin Institute,* Vol. 257 No. 3, 1954.
20. Korfhage, R.R. *Logic and Algorithms.* New York: John Wiley & Sons, Inc., 1966.

21. Marcus, M.P. *Switching Circuits for Engineers*. Reading, MA: Addison-Wesley Publishing Co., 1967.
22. Maroney, G.E. *A Synthesis Technique for Asynchronous Digital Control Networks*. Unpublished M.Sc. Report, Stillwater, OK: Oklahoma State University, 1969.
23. ———— & Fitch, E.C. "A Synthesis Technique for Fluid Logic Control Networks," *Fluidics Quarterly*, Vol. 3 No. 1, p38ff, Jan. 1971.
24. McCloy, D. and Martin, H.R. *Control of Fluid Power*. Chichester, England: Ellis Horwood Ltd., 1980.
25. McCluskey, E.J. *Introduction to the Theory of Switching Circuits*. New York: McGraw-Hill Book Co., 1965.
26. Miller, R.E. *Switching Theory*, 2 vols., New York: John Wiley & Sons, 1965.
27. Moore, E.F., ed. *Sequential Machines: Selected Papers*, Reading, MA: Addison-Wesley Publishing Co., 1964.
28. Ostwald, P.F. *Cost Estimating*, 2d ed., Englewood Cliffs, NJ: Prentice-Hall, Inc., 1984.
29. Perret, R. & David, R. "Synthesis of Sequential Circuits Using Basic Cell Elements," *Proceedings of the JACC*, pp 582–596, 1968.
30. ———— & LeBourgeols, F. "Wired Sequential Control For a Pilot Distillation Plant." *Papers 71-WA/AUT-17 & 18*, presented at ASME Winter Annual Meeting, 1971.
31. Poix, A., Takahashi, Y., & Thal-Larsen, H. "Hazards in Pneumatic Fluidic Circuits." *Paper 68-WA/AUT-18*, presented at ASME Winter Annual Meeting, 1968.
32. Rhyne, V.T. *Fundamentals of Digital Systems Design*. Englewood Cliffs, NJ: Prentice-Hall, Inc., 1973.
33. Roth, C.H., Jr. *Fundamentals of Logic Design*. New York: West Publishing Co., 1985.
34. Shannon, C.E. "A Symbolic Analysis of Relay and Switching Circuits," *Transactions of the AIEE*, Vol. 57, 1938.
35. Wood, P.E., Jr. *Switching Theory*. New York: McGraw-Hill Book Co., 1968.
36. Woods, R.L. *The State Matrix Method for the Synthesis of Digital Logic Systems*. Unpublished M.Sc. Thesis, Stillwater, OK: Oklahoma State University, 1971.
37. ————. "User's Guide for Logic Synthesis Program LOGSYN," *Proceedings of the BFPR*, Stillwater, OK: Oklahoma State University, 1970.

PROG: TRUTAB.BAS

```
1 'THIS TRUTH TABLE PROGRAM WAS WRITTEN BY S.B.FRIEDMAN
2 'TO USE THIS PROGRAM, THE INTERACTIVE INSTRUCTIONS SHOULD
  BE FOLLOWED EXACTLY.  ANSWER ALL QUESTIONS IN THE EXACT FORM
  REQUESTED.
10 Z=0
20 M=0
30 N=0
40 FLAG=0
50 CLS
60 PRINT "INPUT LOGIC FUNCTION AS '240 Z=fn(A, B, ... , H)'
   THEN <RUN 70>":STOP
70 INPUT "ENTER THE TRUTH VALUE OF Z: 1=TRUE, 0=FALSE, DC=DON'T
   CARE";ZZ$:PRINT
80 INPUT "NO. OF VARIABLES: ";NN:PRINT:PRINT"   A  B  C  D  E  F
   G  H    Z":PRINT
90 LPRINT: LPRINT TAB(20)"  A  B  C  D  E  F  G  H    Z":LPRINT
100 FOR A=0 TO 1
110 FOR B=0 TO 1
120 IF NN<3 THEN 240
130 FOR C=0 TO 1
140 IF NN<4 THEN 240
150 FOR D=0 TO 1
160 IF NN<5 THEN 240
170 FOR E=0 TO 1
180 IF NN<6 THEN 240
190 FOR F=0 TO 1
200 IF NN<7 THEN 240
210 FOR G=0 TO 1
220 IF NN<8 THEN 240
230 FOR H=0 TO 1
240                           'LOGIC FUNCTION GOES HERE
250 IF (FLAG=0 AND N>Z) THEN N=Z
260 IF FLAG=1 THEN 460
270 IF (FLAG=2 AND (Z+ABS(N)>0)) THEN M=M+1
280 IF NN<3 THEN 400
290 IF NN<4 THEN 390
300 IF NN<5 THEN 380
310 IF NN<6 THEN 370
320 IF NN<7 THEN 360
330 IF NN<8 THEN 350
340 NEXT H
350 NEXT G
360 NEXT F
370 NEXT E
380 NEXT D
390 NEXT C
400 NEXT B
410 NEXT A
420 IF FLAG=0 THEN FLAG=2: GOTO 100
430 IF FLAG=1 THEN PRINT"--------------------------":LPRINT:
    LPRINT: LPRINT: LLIST 240: END
440 LET FLAG=1
450 GOTO 100
460 Z=Z+ABS(N)
470 IF ZZ$="1" THEN GOTO 800
480 IF ZZ$="0" THEN GOTO 880
490 IF (M<=128 AND NN=8 AND Z=1) THEN 640
500 IF (M>128 AND NN=8 AND Z=0) THEN 640
510 IF (M<=64 AND NN=7 AND Z=1) THEN 660
520 IF (M>64 AND NN=7 AND Z=0) THEN 660
530 IF (M<=32 AND NN=6 AND Z=1) THEN 680
540 IF (M>32 AND NN=6 AND Z=0) THEN 680
550 IF (M<=16 AND NN=5 AND Z=1) THEN 700
560 IF (M>16 AND NN=5 AND Z=0) THEN 700
570 IF (M<=8 AND NN=4 AND Z=1) THEN 720
580 IF (M>8 AND NN=4 AND Z=0) THEN 720
590 IF (M<=4 AND NN=3 AND Z=1) THEN 740
600 IF (M>4 AND NN=3 AND Z=0) THEN 740
610 IF (M<=2 AND NN=2 AND Z=1) THEN 760
620 IF (M>2 AND NN=2 AND Z=0) THEN 760
630 GOTO 280
640 PRINT A;B;C;D;E;F;G;H;"   ";Z
650 LPRINT TAB(20) A;B;C;D;E;F;G;H;"   ";Z: GOTO 280
660 PRINT A;B;C;D;E;F;G;"   ";Z
670 LPRINT TAB(20) A;B;C;D;E;F;G;"   ";Z: GOTO 280
680 PRINT A;B;C;D;E;F;"   ";Z
690 LPRINT TAB(20) A;B;C;D;E;F;"   ";Z: GOTO 280
700 PRINT A;B;C;D;E;"   ";Z
710 LPRINT TAB(20) A;B;C;D;E;"   ";Z: GOTO 280
720 PRINT A;B;C;D;"   ";Z
730 LPRINT TAB(20) A;B;C;D;"   ";Z: GOTO 280
740 PRINT A;B;C;"   ";Z
750 LPRINT TAB(20) A;B;C;"   ";Z: GOTO 280
760 PRINT A;B;"   ";Z
770 LPRINT TAB(20) A;B;"   ";Z: GOTO 280
780 PRINT "------------------"
790 END
800 IF (NN=8 AND Z=1) THEN 640
810 IF (NN=7 AND Z=1) THEN 660
820 IF (NN=6 AND Z=1) THEN 680
830 IF (NN=5 AND Z=1) THEN 700
840 IF (NN=4 AND Z=1) THEN 720
850 IF (NN=3 AND Z=1) THEN 740
860 IF (NN=2 AND Z=1) THEN 760
870 GOTO 280
880 IF (NN=8 AND Z=0) THEN 640
890 IF (NN=7 AND Z=0) THEN 660
900 IF (NN=6 AND Z=0) THEN 680
910 IF (NN=5 AND Z=0) THEN 700
920 IF (NN=4 AND Z=0) THEN 720
930 IF (NN=3 AND Z=0) THEN 740
940 IF (NN=2 AND Z=0) THEN 760
950 GOTO 280
```

PROG: TRUTAB1.BAS (NO PRINTER OPTION)

```
1 'THIS TRUTH TABLE PROGRAM WAS WRITTEN BY S.B.FRIEDMAN
2 'TO USE THIS PROGRAM, THE INTERACTIVE INSTRUCTIONS SHOULD
  BE FOLLOWED EXACTLY. ANSWER ALL QUESTIONS IN THE EXACT
  FORM REQUESTED.
10 Z=0
20 M=0
30 N=0
40 FLAG=0
50 CLS
60 PRINT "INPUT LOGIC FUNCTION AS '240 Z=fn(A, B, ... , H)' THEN
   (RUN 70)":STOP
70 INPUT "ENTER THE TRUTH VALUE OF Z: 1=TRUE, 0=FALSE,
   DC=DON'T CARE";ZZ$:PRINT
80 INPUT "NO. OF VARIABLES: ";NN:PRINT:PRINT" A  B  C  D  E  F
   G
   H  Z":PRINT
100 FOR A=0 TO 1
110 FOR B=0 TO 1
120 IF NN<3 THEN 240
130 FOR C=0 TO 1
140 IF NN<4 THEN 240
150 FOR D=0 TO 1
160 IF NN<5 THEN 240
170 FOR E=0 TO 1
180 IF NN<6 THEN 240
190 FOR F=0 TO 1
200 IF NN<7 THEN 240
210 FOR G=0 TO 1
220 IF NN<8 THEN 240
230 FOR H=0 TO 1
240                          'THE LOGIC FUNCTION GOES
HERE
250 IF (FLAG=0 AND N>Z) THEN N=Z
260 IF FLAG=1 THEN 460
270 IF (FLAG=2 AND (Z+ABS(N)>0)) THEN M=M+1
280 IF NN<3 THEN 400
290 IF NN<4 THEN 390
300 IF NN<5 THEN 380
310 IF NN<6 THEN 370
320 IF NN<7 THEN 360
330 IF NN<8 THEN 350
340 NEXT H
350 NEXT G
360 NEXT F
370 NEXT E
380 NEXT D
390 NEXT C
400 NEXT B
410 NEXT A
420 IF FLAG=0 THEN FLAG=2: GOTO 100
430 IF FLAG=1 THEN PRINT"----------------------------": END
440 LET FLAG=1
450 GOTO 100
460 Z=Z+ABS(N)
470 IF ZZ$="1" THEN GOTO 800
480 IF ZZ$="0" THEN GOTO 880
490 IF (M<=128 AND NN=8 AND Z=1) THEN 640
500 IF (M>128 AND NN=8 AND Z=0) THEN 640
510 IF (M<=64 AND NN=7 AND Z=1) THEN 660
520 IF (M>64 AND NN=7 AND Z=0) THEN 660
530 IF (M<=32 AND NN=6 AND Z=1) THEN 680
540 IF (M>32 AND NN=6 AND Z=0) THEN 680
550 IF (M<=16 AND NN=5 AND Z=1) THEN 700
560 IF (M>16 AND NN=5 AND Z=0) THEN 700
570 IF (M<=8 AND NN=4 AND Z=1) THEN 720
580 IF (M>8 AND NN=4 AND Z=0) THEN 720
590 IF (M<=4 AND NN=3 AND Z=1) THEN 740
600 IF (M>4 AND NN=3 AND Z=0) THEN 740
610 IF (M<=2 AND NN=2 AND Z=1) THEN 760
620 IF (M>2 AND NN=2 AND Z=0) THEN 760
630 GOTO 280
640 PRINT A;B;C;D;E;F;G;H;"   ";Z
650 GOTO 280
660 PRINT A;B;C;D;E;F;G;"      ";Z
670 GOTO 280
680 PRINT A;B;C;D;E;F;"         ";Z
690 GOTO 280
700 PRINT A;B;C;D;E;"            ";Z
710 GOTO 280
720 PRINT A;B;C;D;"               ";Z
730 GOTO 280
740 PRINT A;B;C;"                  ";Z
750 GOTO 280
760 PRINT A;B;"                     ";Z
770 GOTO 280
780 PRINT "------------------"
790 END
800 IF (NN=8 AND Z=1) THEN 640
810 IF (NN=7 AND Z=1) THEN 660
820 IF (NN=6 AND Z=1) THEN 680
830 IF (NN=5 AND Z=1) THEN 700
840 IF (NN=4 AND Z=1) THEN 720
850 IF (NN=3 AND Z=1) THEN 740
860 IF (NN=2 AND Z=1) THEN 760
870 GOTO 280
880 IF (NN=8 AND Z=0) THEN 640
890 IF (NN=7 AND Z=0) THEN 660
900 IF (NN=6 AND Z=0) THEN 680
910 IF (NN=5 AND Z=0) THEN 700
920 IF (NN=4 AND Z=0) THEN 720
930 IF (NN=3 AND Z=0) THEN 740
940 IF (NN=2 AND Z=0) THEN 760
950 GOTO 280
```

PROG:RUNPRIMP.BAT

```
echo off
cls
CONV2 PROB.IN
CLS
PRIMP PROB.IN ANS.OUT
ERASE PROB.IN
TYPE ANS.OUT
```

PROG: CONV2.FOR

```
        CHARACTER TERM(512,10),X0,X1,TABLE(10),DCA
        INTEGER NET,NLIT,NMINT,START,PLACE,NACT,PRES
        INTEGER FTERM(8000),BIN
        DO 5 I=1,8000
          FTERM(I)=0
5       CONTINUE
        X0='0'
        X1='1'
        TABLE(1)='A'
        TABLE(2)='B'
        TABLE(3)='C'
        TABLE(4)='D'
        TABLE(5)='E'
        TABLE(6)='F'
        TABLE(7)='G'
        TABLE(8)='H'
        TABLE(9)='I'
        TABLE(10)='J'

        WRITE (*,300)
        READ (*,100) NLIT
        WRITE (*,310)
        READ (*,110) NMINT
        WRITE (*,320)

        DO 10 I=1,NMINT
          WRITE (*,330) I
          DO 20 J=1,NLIT
            WRITE (*,340) TABLE(J)
            READ (*,120) TERM(I,J)
20        CONTINUE
10      CONTINUE

        NACT=NMINT
        START=1
        PLACE=NMINT+1
        NET=0

        DO 30 I=1,NLIT
          DO 40 J=START,START+NACT-1
            IF (TERM(J,I) .EQ. '-') THEN
              DO 50 K=1,I-1
                TERM(PLACE,K)=TERM(J,K)
                TERM(PLACE+1,K)=TERM(J,K)
50            CONTINUE

              TERM(PLACE,I)=X0
              PLACE=PLACE+1
              TERM(PLACE,I)=X1
              PLACE=PLACE+1

              DO 60 K=I+1,NLIT
                TERM(PLACE-1,K)=TERM(J,K)
                TERM(PLACE-2,K)=TERM(J,K)
60            CONTINUE
              NET=NET+1
            ELSE
              DO 70 K=1,NLIT
                TERM(PLACE,K)=TERM(J,K)
70            CONTINUE
              PLACE=PLACE+1
            ENDIF
40        CONTINUE

          START=START+NACT
          NACT=NACT+NET
          NET=0
30      CONTINUE

        DO 80 K=START,START+NACT-1
          PRES=0
          DO 90 I=1,NLIT
            BIN=(2**(NLIT-I))
            IF (TERM(K,I) .EQ. '1') PRES=PRES+BIN
90        CONTINUE
          FTERM(PRES+1)=1
80      CONTINUE

        WRITE (*,400)
400     FORMAT (' DO YOU WISH TO ENTER DONT CARES? (Y)ES OR (N)O'/)
        READ (*,410) DCA
410     FORMAT (A1)
        IF (DCA .EQ. 'Y') THEN

          WRITE (*,310)
          READ (*,110) NMINT
          WRITE (*,320)

          DO 101 I=1,NMINT
            WRITE (*,330) I
            DO 201 J=1,NLIT
              WRITE (*,340) TABLE(J)
              READ (*,120) TERM(I,J)
201         CONTINUE
101       CONTINUE

          NACT=NMINT
          START=1
          PLACE=NMINT+1
          NET=0

          DO 301 I=1,NLIT
            DO 401 J=START,START+NACT-1
              IF (TERM(J,I) .EQ. '-') THEN
                DO 501 K=1,I-1
                  TERM(PLACE,K)=TERM(J,K)
                  TERM(PLACE+1,K)=TERM(J,K)
501             CONTINUE

                TERM(PLACE,I)=X0
                PLACE=PLACE+1
                TERM(PLACE,I)=X1
                PLACE=PLACE+1

                DO 601 K=I+1,NLIT
                  TERM(PLACE-1,K)=TERM(J,K)
                  TERM(PLACE-2,K)=TERM(J,K)
```

PROG: CONV2.FOR

```
601        CONTINUE
           NET=NET+1
        ELSE
           DO 701 K=1,NLIT
              TERM(PLACE,K)=TERM(J,K)
701        CONTINUE
           PLACE=PLACE+1
        ENDIF
401     CONTINUE

        START=START+NACT
        NACT=NACT+NET
        NET=0

301     CONTINUE

        DO 801 K=START,START+NACT-1
           PRES=0
           DO 901 I=1,NLIT
              BIN=(2**(NLIT-I))
              IF (TERM(K,I) .EQ. '1') PRES=PRES+BIN
901        CONTINUE
           FTERM(PRES+1)=-1
801     CONTINUE
        ENDIF

        IF (DCA .NE. 'Y') THEN
        WRITE (*,409)
409     FORMAT (' DO YOU WISH TO ENTER OFF TERMS? (Y)ES OR (N)O'/)
        READ (*,411) DCA
411     FORMAT (A1)
        IF (DCA .EQ. 'Y') THEN

        DO 11 I=1,8000
           IF (FTERM(I) .EQ. 0) FTERM(I)=-1
11      CONTINUE

        WRITE (*,310)
        READ (*,110) NMINT
        WRITE (*,320)

        DO 102 I=1,NMINT
           WRITE (*,330) I
           DO 202 J=1,NLIT
              WRITE (*,340) TABLE(J)
              READ (*,120) TERM(I,J)
202        CONTINUE
102     CONTINUE

        NACT=NMINT
        START=1
        PLACE=NMINT+1
        NET=0

        DO 302 I=1,NLIT
           DO 402 J=START,START+NACT-1
              IF (TERM(J,I) .EQ. '-') THEN
                 DO 502 K=1,I-1
                    TERM(PLACE,K)=TERM(J,K)
                    TERM(PLACE+1,K)=TERM(J,K)
502              CONTINUE

                 TERM(PLACE,I)=X0
                 PLACE=PLACE+1
                 TERM(PLACE,I)=X1
                 PLACE=PLACE+1

                 DO 602 K=I+1,NLIT
                    TERM(PLACE-1,K)=TERM(J,K)
                    TERM(PLACE-2,K)=TERM(J,K)
602              CONTINUE
                 NET=NET+1
              ELSE
                 DO 702 K=1,NLIT
                    TERM(PLACE,K)=TERM(J,K)
702              CONTINUE
                 PLACE=PLACE+1
              ENDIF
402        CONTINUE

           START=START+NACT
           NACT=NACT+NET
           NET=0

302     CONTINUE

        DO 802 K=START,START+NACT-1
           PRES=0
           DO 902 I=1,NLIT
              BIN=(2**(NLIT-I))
              IF (TERM(K,I) .EQ. '1') PRES=PRES+BIN
902        CONTINUE
           FTERM(PRES+1)=0
802     CONTINUE
        ENDIF
        ENDIF

        WRITE (6,200) NLIT
        WRITE (6,210) (FTERM(K),K=1,2**NLIT)

100     FORMAT (I2)
110     FORMAT (I2)
120     FORMAT (A1)
200     FORMAT (I2)
210     FORMAT (40(I2))
300     FORMAT (' ENTER THE NUMBER OF LITERALS')
310     FORMAT (' ENTER THE NUMBER OF MINTERMS')
320     FORMAT (/' IF A LITERAL IS PRESENT IN THE MINTERM, ENTER A 1'/
       * 'IF IT IS COMPLEMENT IS PRESENT, ENTER A 0'/
       * 'IF IT IS NOT PRESENT, ENTER A -'/)
330     FORMAT (/' MINTERM NO. ',I2)
340     FORMAT (1X,A1,'?')

        STOP
        END
```

PROG: PRIMP.FOR

```
      CHARACTER*4 X(10),X1,X2,X3,XE(3),XC(3),Y(10)
      INTEGER C(1024),MINT(1024),DIF
     *,NUM(1024),PRIMI(512),PRIMJ(512),COST(512),T(512),
     *TABI(1024),TABLE(6000)
      REAL*4 TX(1024),TY(512),CIST
      X1=' X '
      X2=' 1 '
      X3=' 0 '
      XE(1)='ESSE'
      XE(2)='NTIA'
      XE(3)='L   '
      XC(1)='CHOS'
      XC(2)='EN  '
      XC(3)='    '
      Y(1)='A'
      Y(2)='B'
      Y(3)='C'
      Y(4)='D'
      Y(5)='E'
      Y(6)='F'
      Y(7)='G'
      Y(8)='H'
      Y(9)='I'
      Y(10)='J'
      READ (5,150) N
150   FORMAT (I2)
      N2=2**N
      READ (5,151) (C(I),I=1,N2)
151   FORMAT (40I2)
      A2=ALOG(2.0)
      NT=2**N
      MK=0
      M=0
      DO 125 I=1,NT
      TABI(I)=1
      IF (C(I) .EQ. 0) GO TO 125
      M=M+1
      MINT(M)=I-1
125   CONTINUE
      WRITE (6,995) N
      MP=0
      DO 3 I=1,NT
3     NUM(I)=I-1
      WRITE(6,700) (NUM(I),C(I),I=1,NT)
C
C     SOLVE FOR PRIME IMPLICANTS
C
      DO 4 I=1,M
      II=MINT(I)
      MI=M-I+1
      DO 5 J=1,MI
      NN=0
      JJ=MINT(M-J+1)
      IF (IAND(II,JJ) .NE. II) GO TO 5
      CIST=1.0
      DIF=JJ-II
      IF (C(II+1) .GT. 0) NN=NN+1
      IF (C(JJ+1) .GT. 0) NN=NN+1
      IF (DIF .EQ. 0) GO TO 9
      ICNT=1
      NUM(1) = II
      DO 6 K=1,N
      IZ=IAND(2**(K-1),DIF)
      IF (IZ .EQ. 0) GO TO 6
      IC2=ICNT
      DO 8 L=1,IC2
      ICNT=ICNT+1
      NUM(ICNT)=NUM(L)+IZ
      LL=NUM(ICNT)+1
      IF (C(LL))8,5,30
30    NN=NN+1
8     CONTINUE
6     CONTINUE
      CIST=ICNT+1
9     IF(NN.EQ.0) GO TO 5
      IF (MP .EQ.0) GO TO 10
      DO 11 L=1,MP
      IF (PRIMJ(L) .LT. JJ) GO TO 11
      IF (IAND(PRIMJ(L),JJ) .NE. JJ) GO TO 11
      IF (IAND(PRIMI(L),II) .EQ. PRIMI(L)) GO TO 5
11    CONTINUE
10    MP=MP+1
      PRIMI(MP)=II
      PRIMJ(MP)=JJ
      T(MP)=0
      ICOST=ALOG(CIST)/A2
      COST(MP)=N-ICOST
5     CONTINUE
      IF (C(II+1) .LT. 0) GO TO 4
      TABI(I)=MK+1
      DO 13 L=1,MP
      IF(PRIMJ(L) .LT. II) GO TO 13
      IF (IAND(PRIMI(L),II).NE. PRIMI(L) ) GO TO 13
      IF (IAND(PRIMJ(L),II) .NE. II) GO TO 13
      MK=MK+1
      TABLE(MK)=L
13    CONTINUE
      TABI(I+1)=MK+1
      IC=MK-TABI(I)
      IF(IC .EQ. 0) GO TO 15
      A=IC
      TX(I)=1.0/A
      GO TO 4
C
C     SET ESSENTIAL TERM
C
15    J=TABLE(TABI(I))
      T(J)=-1
      NUM(1)=PRIMI(J)
      LL=NUM(1)+1
      C(LL)=-1
      ICNT=1
      DIF=PRIMJ(J)-PRIMI(J)
      DO 16 K=1,N
      IZ=IAND(2**(K-1),DIF)
      IF (IZ .EQ. 0) GO TO 16
      IC2=ICNT
      DO 17 L=1,IC2
```

PROG: PRIMP.FOR

```
            ICNT=ICNT+1
            NUM(ICNT)=NUM(L)+IZ
            LL=NUM(ICNT)+1
 17         C(LL)=-1
 16      CONTINUE
  4   CONTINUE
C
C     ALL PRIME IMPLICANTS ARE FOUND
C
         LL=0
         DO 27 L=1,M
         IF (C(MINT(L)+1)) 27,27,25
 25      LL=LL+1
         NUM(LL)=L
 27      CONTINUE
         IF (LL .EQ. 0) GO TO 31
C
C     SOLVE CYCLIC CHART
C
         BIX=0.0
         DO 28 I=1,LL
         IF (BIX .GE. TX(NUM(I))) GO TO 28
         BIX =TX(NUM(I))
         II=NUM(I)
 28      CONTINUE
 48      DO 49 I=1,MP
 49      TY(I)=0.0
         DO 50 I=1,LL
         IT=TABI(NUM(I))
         JT=TABI(NUM(I)+1)-1
         DO 50 J=IT,JT
         TY(TABLE(J))=TY(TABLE(J))+TX(NUM(I))
 50      CONTINUE
 55      BIY=0.0
         IT=TABI(II)
         JT=(TABI(II+1)-1)
         DO 60 J=IT,JT
         IF (TY(TABLE(J))-BIY) 60,56,57
 56      IF (BIY .EQ. 0.0) GO TO 60
         IF (COST(JJ) .LE. COST(TABLE(J))) GO TO 60
 57      BIY=TY(TABLE(J))
         JJ=TABLE(J)
 60      CONTINUE
C
C     CHOOSE PRIME IMPLICANT
C
         T(JJ)=1
         BIX=0.0
         DO 65 I=1,LL
         IJ=MINT(NUM(I))
         IF (PRIMJ(JJ) .LT. IJ) GO TO 61
         IF (IAND(PRIMJ(JJ),IJ) .NE. IJ) GO TO 61
         IF (IAND(PRIMI(JJ),IJ) .NE. PRIMI(JJ)) GO TO 61
         TX(NUM(I))=0.0
 61      IF (BIX .GE. TX(NUM(I))) GO TO 65
         BIX=TX(NUM(I))
         II=NUM(I)
 65      CONTINUE
         IF (BIX .NE. 0.0) GO TO 48
```

```
 31      CONTINUE
C
C     WRITE OUTPUT
C
         WRITE (6,900) (Y(K),K=1,N)
         DO 40 J=1,MP
         I3=PRIMI(J)
         K3=PRIMJ(J)-I3
         DO 35 K=1,N
         K1=K-1
         X(N-K1)=X3
         IF (IAND(I3,2**K1) .NE. 0) X(N-K1)=X2
         IF (IAND(K3,2**K1) .NE. 0) X(N-K1)=X1
 35      CONTINUE
         IF (T(J)) 38,37,39
 37      WRITE (6,1003) J,COST(J),(X(I),I=1,N)
         GO TO 40
 38      WRITE (6,1003) J,COST(J),(X(I),I=1,N),XE
         GO TO 40
 39      WRITE (6,1003) J,COST(J),(X(I),I=1,N),XC
 40      CONTINUE
         WRITE (6,2001)
         IF(LL .EQ. 0) STOP
         WRITE (6,1004)
         DO 70 I=1,M
         IT=TABI(I)
         JT=TABI(I+1)-1
         IF (C(MINT(I)+1) .GT. 0) WRITE (6,1002) MINT(I),(TABLE(J)
        *,J=IT,JT)
 70      CONTINUE
         WRITE (6,753)
 995     FORMAT('1',T21,'*** BOOLEAN MINIMIZATION PROGRAM **',
        *///T16,'THIS FUNCTION CONTAINS',I5,'VARIABLES',//
        *T16,'A LISTING OF THE INPUT DATA FOLLOWS',/T16,'TR'
        *,'UE MINITERMS = 1',/T16,'FALSE MINITERMS = 0',/T16,
        *'REDUNDANT MINITERMS (DONT CARES) = -1',/)
 700     FORMAT ('0',(T6,5(I5,' = ',I2,' ')))
 900     FORMAT('0',T6,'THE FOLLOWING IS A LIST OF THE PRIME '
        *,'IMPLICANTS OF THE',/T6,'MINIMIZED FUNCTION.',//
        *T10,'ESSENTIAL PRIME IMPLICANTS ARE SO LABELED, AND'
        *,/T10,'PRIME IMPLICANTS SELECTED FROM A CYCLIC CHA'
        *,'RT ARE',/T10,'LABELED AS CHOSEN.',//5X,'    NO.',
        * 3X,'COST',7X,'PRIME IMPLICANTS',/23X,10A4,/)
 1003    FORMAT(' ',5X,I4,2X,I4,6X,13A4)
 2001    FORMAT(' ',/,5X,'X INDICATES A MISSING VARIABLE, 0 IN'
        *,'DICATES A COMPLEMENTED',/5X,'VARIABLE AND 1 INDIC'
        *,'ATES A TRUE VARIABLE.',/5X,'THE FUNCTION IS REPRE'
        *,'SENTED BY THE SUM OF BOTH THE ',/5X,'ESSENTIAL AND'
        *,' THE CHOSEN PRIME IMPLICANTS.')
 1004    FORMAT('1',/////,10X,'CONSTRAINT TABLE',//T8,'COV'
        *,'ERED',4X,'COVERING PRIME',/T8,'MINTERM',6X,'IMPLI'
        *,'CANTS.')
 1002    FORMAT(' ',T9,I5,T16,I5,I6,9(/,7X,15I6))
 753     FORMAT('0')
         STOP
         END
C
C     FUNCTION IAND WRITTEN BY DAVID MCCARTY
```

```
PROG: PRIMP.FOR
C
C  THIS FUNCTION PERFORMS THE BOOLEAN AND FUNCTION ON THE
C  I1 AND I2. FOR EXAMPLE:
C
C                I1 = 54 = 0 1 1 0 1 1 0
C                I2 = 94 = 1 0 1 1 1 1 0
C       IAND (I1,I2)= 22 = 0 0 1 0 1 1 0
C
       FUNCTION IAND (I1,I2)
       INTEGER BI1(10),BI2(10),TEST,BIN,II1,II2
       II1=I1
       II2=I2
       DO 10 I=1,10
         TEST=2**(10-I)
         IF ((II1-TEST) .GE. 0) THEN
           BI1(I)=1
           II1=II1-TEST
         ELSE
           BI1(I)=0
         ENDIF
         IF ((II2-TEST) .GE. 0) THEN
           BI2(I)=1
           II2=II2-TEST
         ELSE
           BI2(I)=0
         ENDIF
 10    CONTINUE
       IAND=0
       DO 20 L=1,10
         BIN=2**(10-L)
         IF ((BI1(L)+BI2(L)) .EQ.2) IAND=IAND+BIN
 20    CONTINUE
       RETURN
       END
```

Copies of the programs are available from the author at a minimal cost. Please send requests to: SBF Professional Engineers, Inc. 3242 4th Street, Boulder CO 80304–2105

Index

Active components, 159
Addition, 18
Adjacency, graph, 162
   requirement, primary, 168
   requirement, secondary, 168
Advantages: Counters & Steppers, 141
   Counting and Passive MEMORY schemes, 158
Algebraic technique, 35, 51
Alterm, 28
AND, 20
ANS.1, 199
ANS.2, 199
AONI, 84, 105
Asynchronous systems, 14
Attached Logic Diagramming, 76
Attached symbols, 61
Automation, 2, 4
   case study, 186
   successful, 79

Be, 132
Best technique, 182
Binary device, 11
   system, 129

Boolean algebra, 18
Break-even quantity, 193
Bridging, 132
Budgeting, 7

Canonical, 28
Cells, conventional arrangement of, 38
Central Processing Unit, 72
Charges, equipment, 6
   power, 6
   space, 6
   support and engineering, 6
Chosen Set, 166
Classical analysis, 162
Clocks, 70
Code, counting, 44
   Decimal, 131
   Gray, 37, 38
   Left Gray, 37
   Right Gray, 37
Coding, 37, 44
Combinational logic systems, 82
   problems, 104
   synthesis, 104

Computer, 14
  Assisted Design, 196
Complement, 29
Compatibility, 162
Components, 56
Conditioning, 68
Conjunctive, 28
Constraints, 2, 56
Contact bounce, 94
  forms, 132
Control system, 12
Cost, 5, 77, 79, 113
  of being wrong, 98
  components, 3
  extended man-machine, 7
  function, 35
  maintenance and lost-opportunity, 5
  operating, 5
Costing Procedures, 5
Counter controlled automation, 138
Counters, binary, 129, 130, 137
CPU, 72

Decimal system, 129
DELAY-IN, 70
DELAY-OUT, 70
DeMorgan's theorem, 29
Design, Active Memory, 168
  Computer Assisted, 196
  objective, 3
  technique, 83, 88, 182
Deterministic systems, 151
Device efficiency, 78
Diagramming, detached-logic, 73
Disjunctive, 28
Division, 18
Do Nothing Option, 113
Documentation, 77
Drum steppers, 131, 135
Dual, 25, 30

Economic level, appropriate, 3
Economics, automation, 3
Electrical switches, 58
Electro-mechanical, devices, 145
  interlock, 62
  rotary stepping switches, 131
Electronic, 68
Environmental conditions, hazardous, 79
EPC, 116
Equality, 18
Equivalent, 29
  Pairs Chart, 116, 162
  Stable states, 115

Excitation Chart, 174
  Maps, 174
Experiment, 10
Expression minimization, 107
Extended mapping technique, 105
Extraction of a minimal statement, 42

Fabrication Considerations, 4
Fanning-in, 64
Fluidic, 68
  bistable device, 160
  passive MEMORY devices, 145
Fluid Power Systems, 62
FORTRAN, 77

George Boole, 10

Hardware, Active Memory, 158
  automation, 2
  injudicious selection of, 94
  selection, 153
Hazard, 89, 105, 176
  dynamic, 89, 92
  elimination of, 91, 98
  static, 89, 90, 92
  static function, 92
  static logic, 91
Hazard-free circuits, 94
Hazards, combinational, 89
  reintroduction of, 93
  sequential, 167, 177, 181
Hierarchy, 24

IF-THEN, 106, 108, 109
Iff, 29
Impact modulator, 69
Implementation, 77, 98
IMPLICATION, 106
Incomplete specification, 49
INHIBIT, 30, 31, 56, 61, 62, 66, 68
Input power, 78
Interlock, 61
Inversion, 18

Justification, 5

Ladder diagram, 72, 73
Literal, 28
Logic, Basic statement of, 10
  diagram, 86
  essential, 126
  formal, 10
  implementation schemes, 84
  simplification, 34
  specification chart, 111

Index

system, 56, 77, 79
symbols, 83
Logic systems, combinational, 82
  counting systems, 128
  sequential, 126, 144, 158
    active, 158
    deterministic, 151
    passive, 144
    stochastic, 148
Logical, automation systems, 10
  reasoning, 10
  synthesis, 13
LSC, 111
  purpose of, 115
  Rules for creating, 112

Machine automation, 79
Maintainability, 62, 73
Map, Karnaugh, 38, 39
Mapping method, 38, 51
Maxterm, 40
Mechanical logic elements, 57
MEMORY, 21, 56, 61, 66, 68, 76
Merging Graph, 117, 162, 164
Microprocessor, 14
Minimal, 34
Minimal-cost multiple output system, 50
  system, 44
  Set, 166
Minimization techniques, 44
Minimum sum expression, 48
Minterm, 40
Moving part logic, 62
MPL, 62
MRSC, 162
Multiplication, 18
Multivibrators, 70

NAND, 20, 30
NB, 132
Next State Table, 119
Noise, 69, 73
Nonanalytic elements of design, 88
Non-bridging, 132
Noncanonical expression, 49
NOR, 21, 31
NOT, 19, 57, 66, 69
NOT INHIBIT, 106

ONE-SHOT, 70, 71
Operational Flow Chart, 166, 170, 173, 174
Operations, 18
Optimal system, 56
Optimality, 34

Optimum engineering design, 158
Optional states, 42
  value, 38
OR, 20, 57, 64
Ordinal integer number, 128
Output, Assignment Table, 179
  equations, 150, 181
  states, 162, 179, 181

Passive memory, 144
  hardware, 144
Pathological event, 97
PFT, 109
  Rules for creating, 110
PLC, 71
Postulates, 25
Prepared flow path, 149
Prime implicant, 43, 44, 45, 51, 86, 117
PRIMIMP, 52, 196
Primitive flow table, 109
Probabilistic, 109
Programmable controller, 14, 71
  comparisons to other systems, 72
Programming language, 73, 77
Programming PLCs, 73
Proofs, 43
PRTANS, 199
Publications, 56

Quine-McCluskey method, 45

Race, 167
Reduced Specification Chart, 117, 120
Relay, general purpose, 59
  long frame telephone type, 60
  reed, 60
Reliability, 34, 60, 62, 77, 113
Reprogramming, 73
RESET, 21
Response time, 34
Rotary, solenoids, 131, 134
  stepper, 131, 139
  Stepping Switch, 60, 132
RSC, 117

Scalar, 18
Scoring technique, 35
Selected sets, 120
Sequential design, 104
SET, 25
SET/RESET, equations, 175, 176
  switching conditions, 149
Signal treatment, 114
Source state, 148, 152
SPDT, 58, 62

SPSTNC, 58
SPSTNO, 58
Stable state, 109
State, signals, 150
  Table, 117, 162
Steppers, 129
Stochastic, 109, 148
Subtraction, 18
Symbols, 82
Synchronous systems, 14
System history, 111
Switches, 58
Switching nomenclature, 132

T Flip-Flop, 129
Term, 28
TFF, 130
Theorems, 25, 27
  Proofs, 26
Time, constraints, 35
  synchronization, 14

Timing, 14
  requirements, 89
Transitional state, 162, 167, 180
Triggered input, 129
TRUTAB, 49, 199
Truth Tables, 19, 26
Turbulence amplifier, 69

Universal operators, 30

Valves, ball, 63
  diaphragm, 63
  poppet, 62
  spool, four-way, 63
  spool, three-way, 62
Variables, 18, 28
  independent, 29

Wall attachment element, 160
Wave shaping, 70

XNOR, 21
XOR, 20